FOR THE BIRDS

NATURE, SOCIETY, AND CULTURE

Scott Frickel, Series Editor

A sophisticated and wide-ranging sociological literature analyzing nature-society-culture interactions has blossomed in recent decades. This book series provides a platform for showcasing the best of that scholarship: carefully crafted empirical studies of socio-environmental change and the effects such change has on ecosystems, social institutions, historical processes, and cultural practices.

The series aims for topical and theoretical breadth. Anchored in sociological analyses of the environment, Nature, Society, and Culture is home to studies employing a range of disciplinary and interdisciplinary perspectives and investigating the pressing socio-environmental questions of our time—from environmental inequality and risk, to the science and politics of climate change and serial disaster, to the environmental causes and consequences of urbanization and war making, and beyond.

Available titles in the Nature, Society, and Culture series:

Diane C. Bates, *Superstorm Sandy: The Inevitable Destruction and Reconstruction of the Jersey Shore*
Elizabeth Cherry, *For the Birds: Protecting Wildlife through the Naturalist Gaze*
Cody Ferguson, *This Is Our Land: Grassroots Environmentalism in the Late Twentieth Century*
Aya H. Kimura and Abby Kinchy, *Science by the People: Participation, Power, and the Politics of Environmental Knowledge*
Anthony B. Ladd, ed., *Fractured Communities: Risk, Impacts, and Protest against Hydraulic Fracking in U.S. Shale Regions*
Stefano B. Longo, Rebecca Clausen, and Brett Clark, *The Tragedy of the Commodity: Oceans, Fisheries, and Aquaculture*
Stephanie A. Malin, *The Price of Nuclear Power: Uranium Communities and Environmental Justice*
Kari Marie Norgaard, *Salmon Feeds Our People: Colonialism, Nature, and Social Action*
Chelsea Schelly, *Dwelling in Resistance: Living with Alternative Technologies in America*
Diane Sicotte, *From Workshop to Waste Magnet: Environmental Inequality in the Philadelphia Region*
Sainath Suryanarayanan and Daniel Lee Kleinman, *Vanishing Bees: Science, Politics, and Honeybee Health*

FOR THE BIRDS

Protecting Wildlife through the Naturalist Gaze

ELIZABETH CHERRY

RUTGERS UNIVERSITY PRESS
New Brunswick, Camden, and Newark, New Jersey, and London

Library of Congress Cataloging-in-Publication Data

Names: Cherry, Elizabeth Regan, 1977– author.
Title: For the birds : protecting wildlife through the naturalist gaze /
Elizabeth Cherry.
Description: New Brunswick, New Jersey : Rutgers University Press, 2019. |
Includes bibliographical references and index.
Identifiers: LCCN 2019000963 | ISBN 9781978801059 (pbk.) |
ISBN 9781978801066 (cloth)
Subjects: LCSH: Bird watching. | Bird watching—Political aspects. |
Birds—Conservation. | Wildlife conservation.
Classification: LCC QL677.5 .C44 2019 | DDC 598.072/34—dc23
LC record available at https://lccn.loc.gov/2019000963

A British Cataloging-in-Publication record for this book is available from the British Library.

♾ The paper used in this publication meets the requirements of the American National Standard for Information Sciences—Permanence of Paper for Printed Library Materials, ANSI Z39.48-1992.

www.rutgersuniversitypress.org

Manufactured in the United States of America

CONTENTS

FOR THE BIRDS

INTRODUCTION

THE NATURALIST WHOSE name graces the oldest, largest, and best-known bird conservation organization in the United States regularly killed birds. John James Audubon's *Missouri River Journals* frequently include passages such as "Killed a Catbird, Water-thrush, seventeen Parrakeets, a Yellow Chat, a new Finch, and very curious, two White-throated Finches, one White-crown, a Yellow-rump Warbler, a Gray Squirrel, a Loon, and two Rough-winged Swallows."[1] And that was a slow day. Audubon killed birds for ornithological study and painting, and when he wrote these journals in 1853, naturalists had not yet fathomed anthropogenic (human-originating) causes of extinction.[2] Audubon hated to see birds' deaths wasted and wanted to make certain that any birds he killed died for a good cause: "We saw a Wild Goose running on the shore, and it was killed by Bell; but our captain did not stop to pick it up, and I was sorry to see the poor bird dead, uselessly."[3]

Audubon's journals, with their voluminous lists of assorted bird species, resemble modern-day birders' species lists from field trips with local chapters of the Audubon Society. Now, groups of birders go out on walks, record the numbers and types of bird species they see, and then submit their lists to eBird, an online data repository, to be used for citizen science and wildlife conservation efforts. His namesake society takes up Audubon's love of birds and his thrill of seeing birds in their natural habitats. But contemporary birders, especially those associated with the Audubon Society, hold conservation as a central objective of their activities.

This book examines contemporary birders and their focus on conservation. Birding means more than simply watching birds; it encompasses training one's senses to pay close attention to all of the sights and sounds in nature. In this book, I argue that birders learn to view wildlife in a particular way, which I call the "naturalist gaze." The naturalist gaze provides birders with a systematic understanding of humans' and wild animals' intertwined places in a shared ecosystem. This sensibility, in turn, structures birders' environmental advocacy in the form of citizen science projects and wildlife conservation. The naturalist gaze connects the minute and mundane aspects of birding, such as the lessons birders get on nature walks, to the large and consequential, such as the movement to protect the environment from anthropogenic climate change. This deep understanding of nature and all of its elements creates a way of interacting with the natural world that differentiates birding from other ways of interacting with animals or engaging in nature-based hobbies. As a result, birders should be recognized as important allies in the larger environmental conservation movement.

HISTORY OF BIRDING AND CONSERVATION

When naturalists began studying species for categorization in the mid-1700s, they collected large quantities of dead specimens for study. They didn't concern themselves with conservation because they didn't realize that any particular species could die out if they continued killing animals by the hundreds, or even thousands. Collectors traveled the world, bringing back species for scientific study and for exhibit as entertainment. In 1774, collector Ashton Lever took over twelve rooms of a former royal palace in Leicester Square in London and charged admission to visitors wishing to see his collection of exotic species. As he modestly described it, "I am at this Time, SOLE POSSESSOR OF THE FIRST MUSEUM IN THE UNIVERSE."[4]

John James Audubon's practice of killing birds for study and painting was not unique. In fact, the technique of shooting birds for ornithological study was so widespread that field guides of the nineteenth century provided guidance on how to use lightweight shot to merely stun birds, and how to kill the birds with their bare hands so as to preserve the specimen.[5] Luckily, birding has changed since that time, and conservation groups bearing Audubon's name began to work on protecting birds. Thirty-five years after Audubon's death, the naturalist George Bird Grinnell founded the first "Audubon Society" in 1886. Grinnell intended for the Audubon Society to serve as the popular branch (i.e., open to those outside of the American Ornithologists' Union) of the growing bird protection movement.[6] One of the first concerted campaigns associated with Audubon focused on the use of bird feathers—and whole birds—in women's hats. Birds and bird feathers had featured prominently in hats for centuries, but the trend grew throughout the 1800s. The killing of birds for the feather trade devastated bird populations in the 1880s and 1890s, and the women who founded local chapters of the Audubon Society banded together to oppose the use of birds for fashion: "I believe it is woman's sacred mission to be the conservator of beauty and not its destroyer," said Margaret T. Olmstead, member of an Iowa club.[7]

The women in these birding clubs set to work on a variety of strategies and tactics to end the use of birds and their feathers in women's hats. In addition to making moral arguments against the practice, these women created consumer lists of milliners who sold hats without feathers. The men in these birding clubs used their access to other areas of the public sphere, lobbying legislators and writing letters to the editor.[8] Their work was successful. Women stopped buying hats with feathers, Congress passed the Lacey Act in 1900, which prohibited the trade of ill-gotten wildlife, and local chapters of the Audubon Society across the United States thoroughly embraced conservation as a key element of their mission.

That same year, in 1900, Frank Chapman proposed the first Christmas Bird Count as an alternative to traditional Christmas Day hunting competitions.[9] Chapman, editor of *Bird Lore* (the precursor to *Audubon* magazine), urged readers to take a census of all the birds in their area and report them to the magazine. He hoped to capitalize on the same competitive spirit that pushed hunters to publicize their numbers of kills in sportsmen's journals. The count began modestly but soon grew into an annual tradition for Audubon Societies and their members. The Christmas Bird Count remains one of the largest and longest-running citizen science projects in the world, and birders' regular participation in this and other citizen science initiatives provides valuable data for conservation scientists. Birding is now synonymous with conservation.

CONTEMPORARY BIRDERS: STEREOTYPES, DEMOGRAPHICS, AND MOTIVATIONS

People often stereotype birders.[10] One stereotype is the competitive birder, who travels the world in search of a rare species to add to her "life list," or the complete list of species a birder has seen in her lifetime. This stereotype is rooted in real life, in the well-publicized stories of birders pursuing a "big year," an attempt to see as many different species of birds as possible in one year. Birders often publish memoirs depicting their big year attempts, and newspapers regularly feature stories of birders flying to faraway destinations to see rare birds, even chartering private flights.[11] This competitive side of birding is the stuff of legend and of comedy. The 2001 feature film *The Big Year* follows comedians Steve Martin, Jack Black, and Owen Wilson on their own (fictional) competitive big year. And comedian Jon Stewart, on his satirical news show *The Daily Show*, featured a segment on a friendly "big day" competition called the World Series of Birding.[12]

People also hold a stereotypical image of what a birder looks like: light-colored pants with lots of pockets for holding field guides and notebooks, a vest with even more pockets, a wide-brimmed hat to protect from the sun, and, of course, lots of gear, including binoculars, a telescope on a tripod, and a camera with yet another tripod. This stereotype is also rooted in real life, since such gear is practical and useful to birders—though most birders would not use all of this gear at the same time. Wearing a hat instead of sunglasses allows for easier use of binoculars, and scopes on tripods permit even closer viewing than binoculars provide. Since the ethical, conservation-minded birders I studied for this book do not attempt to physically get close to birds, using this gear allows them to get as visually close as possible without bothering the bird.

People's image of what a birder "looks like" also includes demographic characteristics. The stereotypical birder is older, white, and highly educated, and has disposable income. This image largely aligns with reality. According to a survey of birders conducted by the U.S. Fish & Wildlife Service (USFWS), age and birding participation are positively correlated: 30 percent of those aged fifty-five and up participate in birding, in contrast to only 6 percent of those aged sixteen to twenty-four.[13] The survey shows the average American birder is fifty-three years old. The U.S. Department of Agriculture's National Survey on Recreation and the Environment project also reveals that

86.3 percent of birders are white and that all minority races other than Native American are underrepresented in birding.[14] The USFWS survey shows that the higher a person's income and education, the more likely they are to be a birder. While stereotypical images of competitive birders are men, 56 percent of all birders are women.[15] The birders I met in the course of my research for this book align with these nationwide demographics and with other surveys of participants at birding festivals.[16]

People also stereotype birders as quirky or weird in some way. This remains true even for people who know birders. Rhonda, one of my participants, shared with a small group at her local Audubon chapter meeting that she had told a friend she was going to the Audubon meeting that evening, to which her friend responded, "Oh, you must be really lonely." Rhonda said she asked her friend, "What are *you* doing tonight?" and her friend replied, "Watching *Frasier* re-runs." Everyone in the group laughed and started joking about how they all felt so lonely. It became the phrase of the evening, and the birders then suggested I name this book "The Loneliness of the Long-Distance Birder." Birders know these stereotypes and joke about them. But they also try to combat some of them.

While the conservation-minded birders I studied keep lists of the birds they see, and while they appreciate and often keep their own life lists, they malign "listers," who run from place to place seeking only to add to their species lists, rather than slowing down and enjoying the rest of the scenery.[17] Wishing to make birding more inclusive, birders emphasize the existing diversity within the birding community, and they try to encourage birding in minority communities. Professor of wildlife and National Audubon Society board member Drew Lanham has written about "birding while black," and he and other birders of color strive to instill a love of birding and nature in youth of color.[18] The National Audubon Society's Diversity, Equity, and Inclusion initiative has partnered with local Lincoln, Nebraska, organization OUTLinc to provide guided bird tours for lesbian, gay, bisexual, and transgender (LGBT) folks and allies, which they call LGBT: Let's Go Birding Together.[19] And, rather than "weird," birders are some of the kindest people I have ever met. Their patience with me not only as an ethnographer but also as a birding novice instilled a love of birds and birding in me, as well. If birders aren't simply embodiments of these stereotypes, who are they?

THE NATURALIST GAZE

This book focuses on how contemporary birders view and interact with the natural world. Nonhuman animals have always played a role in human society. Perhaps their greatest role has been a symbolic one, when humans symbolically separate themselves from nonhuman animals: animals represent nature, and humans represent culture. Throughout this book I use the term "animal" to denote nonhuman animals purely for the sake of readability and clarity.[20] Once humans have symbolically separated ourselves from animals, we then create symbolic boundaries between different "types" of animals. We dote on our beloved companion animals, we eat anonymous farmed animals, we experiment on certain types of animals in labs, and we largely ignore the wild animals who exist "out there" in "the wild." Who are these mysterious wild animals, and what do we know about them?

Much of our sociological attention has focused on the exploitation of wildlife as a social problem, meaning its roots are societal and its effects are widespread. In this realm, sociologists have most extensively studied trophy hunting and poaching of wild animals. In comparison with local subsistence hunters who hunt for food, trophy hunters often travel to hunt purely for sport, such as killing deer with large antlers or going on safari to kill exotic animals.[21] After that, a significant amount of our attention has focused on how people keep wild animals for entertainment in zoos, aquariums, and circuses, or how people keep wild animals for hobby or sport, such as pigeon racing.[22] Sociologists have also studied how humans use wild animals for medical purposes, such as the horseshoe crab, whose blood we use for various biomedical purposes.[23] Theoretically, these studies depict how people project meaning onto natural objects, "see" or "find" themselves in nature, or use nature or animals to better understand themselves. But if we study only how people use and exploit wild animals for their own entertainment, or how people understand their own identities through experiencing nature, we miss the myriad ways that wild animals enrich humans' lives and how humans respect wild animals. With my focus on birders' appreciation of individual birds and their behaviors as well as the ecosystems in which we all fit, I show how people remake their entire outlook on the natural world. I also argue that people need to better understand their impacts on nature. Birding provides the perfect arena to do so.

Birding offers one of the few opportunities for people to observe, and help, wild animals. While people may take whale-watching tours or watch nature documentaries, birding presents the most accessible way to directly observe wild animals in their habitats, and therefore birds provide an ideal entry into wildlife conservation issues. Birding represents one of the most popular hobbies in the United States, but it remains one of the least understood in terms of its relationship to and impact on birds and the environment. According to the USFWS, hunters make up just 6 percent of the U.S. population, whereas birders comprise 20 percent of the U.S. population.[24] Despite its popularity, people think birding is "for the birds," an esoteric or asocial pastime. But birding also represents an environmental hobby that is conducted on behalf of birds, or for the birds' needs.

One reason why people think birding is "for the birds" is because culture has taught us to ignore nature. Birding's emphasis on attention and evaluation thus makes it useful to study. Take, for example, another excerpt from John James Audubon's journals, where he describes his landing on Jestico Island off Cape Breton, and he gives careful attention to each element of the natural world:

> On landing we found it covered with well grown grass sprinkled everywhere with the blossoms of the wild strawberry; the sun shone bright, and the weather was quite pleasant. Robins, Savannah Finches, Song Sparrows, Tawny Thrushes, and the American Redstart were found. The Spotted Sand-piper, *Totanus macularius,* was breeding in the grass, and flew slowly with the common tremor of their wings, uttering their "wheet-wheet-wheet" note, to invite me to follow them. A Raven had a nest and three young in it, one standing near it, the old birds not seen. *Uria troile* and *U. grylle* were breeding in the rocks, and John saw several *Ardea herodias* flying in pairs, also a pair of Red-breasted Mergansers that had glutted themselves with fish so that they were obliged to disgorge before they could fly off. Amongst the plants the wild gooseberry, nearly the size of a green pea, was plentiful, and the black currant, I think of a different species from the one found in Maine. The wind rose and we returned on board. John and the sailors almost killed a Seal with their oars.[25]

Audubon's journals swell with such majestic, thorough, and detailed notes on the species that he observed (or killed). Audubon knew what to pay attention to and how to place it in context, such as differentiating the gooseberry from the currant. Others likely would have passed right by those bushes without even knowing the plants were edible. These differences in attention and evaluation do not stem from any sort of inherent differences between those who study nature and those who do not—those differences are also cultural.

Culture teaches us what to pay attention to and what to ignore. Cognitive sociologists have shown how we learn to pay attention to immediate dangers, for example, and ignore other, less crucial information.[26] Just as the sociological theorist Georg Simmel's urbanites developed a blasé attitude so as not to get overwhelmed with the amount of stimuli in a city, so too have we developed methods of paying attention only to what is most important, and we learn to tune out that which we deem unimportant.[27] The methods of attention we develop for navigating the cultural world easily translate to the attention we give to the natural world. Car horns in a city and birds chirping in the trees all become background noise. Birders, however, learn to pay rapt attention to the natural world. They pay attention to things others might ignore—sights, sounds, shapes, and even habitat, seasons, and food sources. Each of these elements provides birders with clues to the different species of birds they might expect to see in a particular area.

Culture also teaches us how to evaluate what we see. The attention we give—or do not give—relates to the value we assign to the natural world. We have created symbolic boundaries between ourselves and the natural world, and within the natural world. As noted earlier, people have created symbolic boundaries between humans and animals, and among companion, farmed, and wild animals. Birders further create boundaries between different types of birds that correspond to their perceived value to the natural world. For many birders, native species of birds are "good" birds that deserve more attention than the invasive, nonnative species of birds that damage the ecosystem.

Birding's predominant focus on the environment further makes it important for us to understand as a social practice. While we as a society largely ignore nature, we also socially construct nature—we give it meanings. The social construction of nature refers to the various ways that society imbues nature with meaning, and how we do so differently for different types of nature. To many people, remote wildernesses or grand vistas at national parks represent "real nature," whereas we ignore the equally real nature in our own backyards.[28] In contrast to this common social construction of nature as remote and far away, birders conceive of nature as existing everywhere. This is, in part, because birds fly anywhere and everywhere, and birders are always looking for birds. Birds' mobility also provides us with new ways of thinking about what makes a wild animal "wild," in contrast to our primary encounters with wild animals taking place in captivity in zoos, circuses, or aquariums.

Each of these elements, in combination, leads to birders developing what I call the "naturalist gaze." Birders are naturalists who understand various elements of the natural world, and this encyclopedic viewpoint envelops each of those elements. The naturalist gaze describes a way of looking at birds, in combination with people and other elements of an ecosystem, to create a comprehensive perspective on birds in their natural settings. Birders notice elements of the natural world that other people ignore, and in addition, they pay attention to and evaluate them. Birders develop this naturalist gaze by participating in bird walks and learning how to bird, by reading field guides, and by reading natural histories of birds and the environment. Thus, the naturalist gaze is informed by scientific research on birds and the environment. This means birders can look at any element of the natural world and evaluate it for its benefits and utility to birds and to other parts of the ecosystem, which includes people.

Birders show us how to interact with nature and the wild, in natural and wild settings, in ways that protect wild animals and their habitats. Birders thus also understand the effects that people have on birds and other elements of our shared ecosystems. Birders see, through birds, what many other people refuse to see or acknowledge in our environment: birds are indicator species, meaning they indicate environmental distress through their migration, nesting, and mating patterns, and their extinction or endangerment attests to these environmental issues. This knowledge affects how birders conduct birding and wildlife observations ethically, and how they live their own lives, sustainably. The naturalist gaze is self-sustaining: it encourages birders to participate in citizen science as well as wildlife and environmental conservation efforts. Birders thus contribute to the scientific knowledge that informs the naturalist gaze through citizen science. They also turn the naturalist gaze on themselves, and they participate in a variety of conservation efforts to protect our shared ecosystems.

RESEARCH METHODS AND PARTICIPANTS

Beginning in April 2015, and for a duration of three years, I conducted ethnographic research with birders by attending free, public birding walks given by three local Audubon organizations. Most of these walks occurred in the New York metropolitan area and the lower Hudson River Valley region of New York, where I live. These walks all featured a leader and generally took place in local parks. Usually held on weekends and starting shortly after sunrise (typically at 7:00 A.M. in spring and summer and 8:00 A.M. in fall and winter), the walks lasted for about two hours, as the group meandered on a path through the park. The guide pointed out birds and explained their behavior, and participants also helped spot birds. I attended one to three walks per week during spring and fall migration, which are the most popular times for birding and thus when Audubon chapters hold the most walks, and one to three walks per month in winter and summer, when the birding is slower.

The walks typically included five to ten participants, but sometimes as many as twenty to thirty on particularly beautiful days, especially during spring migration. Most of the walk participants presented as white, a little more than half presented as women, and most participants seemed to be retired and/or of retirement age. Sometimes the walks included younger participants—children, teens, and twentysomethings—but for the most part, the walk participants resembled the demographics of the typical birders described earlier.[29]

I introduced myself as an environmental sociologist interested in learning how to bird, and also intending to write about birding as an environmental hobby. On the walks, I took field notes in a small notebook. This was unobtrusive, since many birders take their own notes on the species they see on the walk. I immediately transcribed and expanded my field notes when I returned home from the field. Sometimes I recorded my thoughts into my smartphone's voice memos application as I drove home, then expanded those when I returned home. As my fieldwork progressed, I moved from being an outside spectator toward becoming more of a full participant. As I became more involved in the groups, one local Audubon society asked me to join its board of directors. In keeping with my feminist research ethics, and wanting to give back to these groups, I agreed.[30]

While all of the walks had the same features when it came to looking for and discussing birds, each group had its own local culture.[31] One group emphasized camaraderie and ended each walk with brunch at a local diner. Another group frequently had beginner birders on its walks and featured more education, in comparison with another group, which had more expert birders and included more discussion and debate over identifications. My walks with Audubon chapters in other states confirmed my general findings and my depiction of these local cultures, as other groups featured each of these elements as well.

My fieldwork encompassed more than the guided bird walks. I also attended the Audubon chapters' monthly nature programs, which include some business but mostly consisted of guest speakers who presented on a variety of topics. Some speakers presented travel slide shows from international birding trips, others presented local natural history, and some brought live animals for more child-friendly programming. I presented my research for this book at several local Audubon chapter meetings as well.

My research also included attending a variety of citizen science events, such as Project FeederWatch, the Christmas Bird Count, and even a dragonfly and damselfly count one summer in July. Once I joined the board of a local Audubon chapter, I also attended its monthly meetings. In July 2015, I attended the National Audubon Convention, its biennial conference, and in May 2016, I visited the World Series of Birding held at Cape May, New Jersey, hosted by New Jersey Audubon. Contrary to what the name implies, the event is not as competitive as it sounds. It's a friendly competition—it's a "big day" event for a good cause, raising money for wildlife conservation efforts. I also attended and observed other National Audubon events, such as its annual meeting and board of directors vote, as well as its national town hall teleconferences.

Once I got an idea of the themes that emerged on birding walks, I began conducting in-depth interviews with birders. I interviewed thirty birders from across the United States to learn more about their motivations for birding, and my interviewees resembled the demographics of typical birders described earlier. The interviews covered a series of questions surrounding five main topics: birders' background and typical birding practices, their relationship to nature, their relationship to birds, citizen science, and environmentalism.[32]

OVERVIEW OF THE BOOK

The arc of the book begins with the process of developing the naturalist gaze, then describes the elements of the naturalist gaze itself, and culminates in the outcomes of the naturalist gaze. Chapter 1 asks how birders learn how to "watch" a bird. Anyone can simply look at a bird, and many people can identify common birds without formal training. In this chapter I show how birders learn how to pay attention to things people typically ignore. The informal lessons birders receive on guided birding walks form the basis for the naturalist gaze—bird walk leaders guide participants through the process of learning how to pay close attention to their surroundings to become adept at hearing, finding, and identifying birds through an informed understanding of bird behavior, diet, and habitat. Participants learn basic skills, such as how to use binoculars, and deeper philosophical lessons on birding ethics.

The art historian John Berger asked, "Why look at animals?"[33] In chapter 2, I ask, "How *should* people look at animals?" and I explain the core elements of birders' naturalist gaze. Birders learn to appreciate healthy wild birds in the wild, in their natural habitat, with minimal human intervention, which differentiates the naturalist gaze from other types of gazes that people use to watch animals. The naturalist gaze helps birders understand power relations in watching wildlife. Thus, it informs birders' codes of ethics for birding, and it inspires birders to engage in wildlife and environmental conservation.

Chapter 3 focuses on the mundane aspects of birding in one's own literal or metaphorical backyard, and I ask how birders understand common birds, or the birds they see every day. Walk leaders teach novice birders to appreciate the everyday beauty and wonder in the most common of birds, and more seasoned birders maintain this admiration of common birds as well. Marveling at the beauty and behavior of the common birds they see every day, through the naturalist gaze, helps birders find and appreciate nature everywhere. Why does this perspective matter? Birders appreciate common birds because they know that when common birds thrive, it means an ecosystem is healthy. Birders also use common birds as an entry point to appreciating the natural environment and as a way of teaching children about nature. Creating this wonder and admiration for the mundane aspects of local nature encourages people to protect the natural environment more than does telling stories of the plight of exotic animals people will likely never see in their natural habitats.

Birds' mobility and omnipresence allow people to view wild animals in the wild in different ways than people typically watch wild animals in captivity. In chapter 4, I show how birds' ubiquity helps connect people to nature. By being anywhere and everywhere, birds bring wildness to people, even in built environments like cities. The fact that birders are "always birding," combined with birds' mobility, further develops the naturalist gaze as birders view themselves and wild animals as part of a shared ecosystem. This knowledge inspires birders to create wilderness in their own backyards, to share with birds.

Why are some birds "good" and other birds "bad"? Chapter 5 explores how birders understand and categorize individual bird behaviors and the habits of particular species. Looking at bird behavior through the naturalist gaze, birders evaluate how birds fit into an ecosystem. "Good" birds provide ecosystem benefits, but "bad" birds are invasive, nonnative species that harm the ecosystem for native birds. However, this distinction is not so clear, since birders also consider humans' interventions into birds' natural habitats. By showing how humans are implicated in "bad" birds' behaviors, this chapter shows the relationship among human agency, animal agency, and animal instinct.

The final two chapters of the book explore the environmental consequences of the naturalist gaze: participating in citizen science and environmental conservation efforts. Birders develop the ability to appreciate the particular and the general, including individual birds, entire species of birds, and birds' place in the ecosystem. In chapter 6, I argue that birders' observational skills, combined with the naturalist gaze, motivate birders to participate in citizen science projects and environmental conservation. Citizen science refers to science conducted in part by nonprofessional scientists, and the greatest contributions to citizen science projects throughout the world come from birders. Birders blur the boundaries between work and leisure, as they spend their leisure time collecting data that field researchers get paid to collect. The naturalist gaze helps birders remain rigorous in their observations, since birders share scientists' skepticism about the reliability and validity of citizen science data. These observations provide evidence of the effects of anthropogenic climate change on birds as indicator species, and thus they inform wildlife and environmental conservation efforts.

Finally, chapter 7 connects birding with larger movements for wildlife and environmental conservation. John James Audubon studied birds for his drawings by shooting them by the dozens. Now, the Audubon Society discourages such rampant killing of birds and has grown into one of the largest conservation organizations in the United States. Birders take the conservation mission of Audubon seriously, and they push for structural changes through Audubon's lobbying efforts. The naturalist gaze helps birders see both people and birds as taking part in a larger, shared ecosystem. Birders are thus able to turn the naturalist gaze back on themselves, and they attempt to lessen their own impact on climate change by making changes in their own lifestyles.

I conclude the book by placing birding and the naturalist gaze back in their larger context of the relationships people hold with wildlife and with nature. More than looking back, though, I encourage readers to look ahead to the myriad environmental problems that we face and how readers might contribute to solving those problems. Birding represents one of the few times that people unobtrusively watch wildlife in its natural habitat. My conclusion demonstrates the implications of better understanding this respectful relationship, and it offers new directions for future research in human-animal studies.

1 ❧ BECOMING A BIRDER

HOW DOES SOMEONE become a birder? Anyone can enjoy looking at a bird, and many people can identify several common birds without any formal training. There isn't any particular moment when a person has a sudden identity shift and becomes a birder. "Birder" describes one's interest and skills in observing and identifying birds more than it does one's identity. Therefore, becoming a birder means learning and building one's birding skills. It also means learning to pay attention to things we typically ignore. Culture teaches us to ignore the "background" of our daily lives, but birders learn to pay attention to detail by learning to find birds by sight and by sound. Birders also learn to place all of these details into a larger context, to identify birds. Once birders master these skills, they can enjoy birding as a meditative activity. These abilities to find and identify birds using contextual clues help birders develop the naturalist gaze.

The birders I interviewed first became interested in birding socially through friends, family, or scouting. Some birders said their birding started as a more solo endeavor, where they got interested in birds by virtue of being an "outdoorsy kid" or having a bird feeder in their backyard. No matter how they initially became interested in birds, everyone I interviewed said they learned how to bird by participating in bird walks. When I interviewed Vivian, she told me that she works in a very corporate environment in New York City. But she grew up playing outside all the time and was "always into nature." She and her brother "had snakes, and lizards, and frogs, and fish," and her grandmother kept bird feeders in her yard and taught Vivian and her brother the local birds with her *Golden Guide* book. Once she and her brother grew up, he became interested in birding, and she found a local bird walk for the two of them to attend when he visited her. Vivian said, "I was hooked instantly" and "got really into field trips" after that visit, which reminded her that she "really always liked to be an outdoor kid and being with nature."

Local and regional Audubon chapters, or other birding clubs, play a major role in becoming a birder. People interested in nature often seek out these clubs or organizations as a way of learning how to bird. When I met Mona at the National Audubon Convention, she told me that group walks were crucial for her learning how to bird: "That is where I really learned how to bird. And to go out, and identify the birds, learn about the habitats, learn about the different colorations of birds, where you find them, and how to ID flying hawks. You don't learn that overnight. . . . And that leads to better binoculars, that leads to out of town trips, and that leads to entire bird-based vacations."

Birds weren't always the initial driving force behind seeking out an Audubon chapter. Sometimes, the birders I interviewed sought out new groups to join, but other times, birding came to them via friends or family. I met Melinda, a teacher, at the World Series of Birding, where she told me that a fellow teacher got her interested in birding. Her colleague brought in field guides to show Melinda the birds at the school. Melinda said, "I started thinking, 'I wonder if I have any of those birds at my house?' So, I started looking, and of course, everybody knows a Blue Jay, everybody knows a Cardinal. But a Chickadee, a Nuthatch, those are all birds that I wasn't familiar with. So, when she started pointing those out, I started noticing how they go up, like the Nuthatch, how it goes down to get its food, instead of up, like the other birds."

Other birders I interviewed became interested in birding because of an abiding interest in nature. Many birders were involved in scouts or in other naturalist activities, which led them to birding. The youngest birder I interviewed was Ethan, who at age eighteen was excited about going to Cornell University in the fall, with the hope of studying ornithology. He said he

FIGURE 1. Tufted Titmouse (photograph courtesy of Dave Saunders).

became interested in birding through "herping," or observing amphibians and reptiles, such as snakes and salamanders, which he had been doing since around age twelve. Herping led Ethan to develop a love for ethically observing animals in their natural environment, where "you don't collect them, you don't poach them, and you don't harass them." By the time he got to high school, he had developed a love for owls and other raptors, and he took a field ecology class, in which he took part in a hawk watch survey. There, he met a senior whom he described as a "bird watcher" and who "instructed my passion into birding."

A final subset of birders became interested in birding on their own, by watching birds at the bird feeder. Even if they became interested in birds on their own, through a backyard bird feeder, they still joined birding groups and went out with other birders so they could learn how to identify birds and improve their birding skills. Joy said she started using the bird feeder the previous homeowner happened to leave in the backyard, which led to her getting a camera to take photographs of the birds and trying to identify what she saw. Tom explained to me that it's difficult to learn how to identify birds on one's own. Many of the birders I interviewed said their identification skills vastly improved once they started going out with other birding groups, as Tom did: "I remember the first bird I ever identified myself out of a book. It was a Tufted Titmouse. And it was just so exciting, to see this, you know, I knew seven or eight birds, but this strange-looking little bird, with like a punk rocker's tuft coming up out of its head, with beautiful colors. . . . The first book I bought was the *Audubon's*, the green book with just photographs. And believe me, it's not a book you want to get if you're a beginner." Tom went on to describe how he learned to identify some birds on his own, through this book, but his birding skills vastly improved when he joined a local Audubon chapter and started going on walks with the group. He continued: "Just having people that you could say to, 'What is this? What am I hearing?' And they could tell you, 'This bird sings its song in the spring, it doesn't sing it in this time of year, in the dead of winter.' But yeah, if you're hearing that sound, you're hearing this particular bird. Then I started learning pretty fast." Tom said his birding improved by "leaps and bounds" once he started going out with a group, and the other birders I interviewed told me the same thing. As I show later in the chapter, guided bird walks teach participants how to find and identify birds by sight, sound, and other indications.

LEARNING ATTENTIVENESS THROUGH SOUND

The birders I interviewed told me they prefer the term "birding" to "bird watching" to describe their practice because birding encompasses more than simply looking at birds. Birding describes the active walking around, looking for birds, and identifying them by sight and sound. Most importantly, birders learn to pay attention to sights and sounds that people typically ignore. As birders learn to pay attention to these elements of nature, they become members of what the cognitive sociologist Eviatar Zerubavel calls an "attentional community."[1] Attentional communities learn to notice specific things that are relevant to their particular group. Moreover, they develop the skills to notice elements that people outside of their community may ignore.[2] These attentional skills often come with specialized training in a profession, or, as I show in the case of birders, with the development of skills for their hobby.[3]

Birders learn to pay attention to details such as habitat, beak shape, or the time of year, to help them discern and identify a bird. They likewise learn to ignore unimportant features, such as the noises of squirrels or chipmunks, which can sound like birds. Birders know they need to pay close attention to their surroundings, in the hopes of finding birds. As one birder put it, in a newspaper column: "In sports there is an expression of how important it is to 'show up.' When it comes to wildlife observation, probably the most critical skill is to be observant. Individuals who are not observant miss out on the abundance of nature that is all around them. . . . Being observant results in seeing things that otherwise go unseen."[4] Birders learn observation skills by going on bird walks, which begin with the leader explaining the route, what the group should expect or hope to find, and some tips on finding the birds as they take their walk. During walks, birders receive both explicit and implicit lessons of birding: while learning a bird's song, for example, birders will also learn that birds typically sing during mating season, and thus they learn to give birds ample space during spring and summer so as not to disturb their mating and nesting areas.

Hearing plays an important role in birding. Birders must pay close attention to sounds, since birds' songs or chip notes provide the first indication that a bird is around to potentially be seen. When I first began attending guided bird walks, I was surprised by the extent to which the leaders on the birding walks encouraged new birders to use multiple senses to find birds. Everyone who attended the bird walks was able to both hear and see birds, and the guides said that this combination helps find more birds.

Bill, who leads the weekly walks for one local Audubon chapter, always encourages paying attention to birdsong as a way to find birds. Bill is a retired construction worker with a gruff attitude, thick Brooklyn accent, and wry sense of humor. He and his wife, Sharon, make a great team of bird walk leaders. At the start of one walk, Bill said, "This is a good group, because we have a lot of eyes and a lot of ears, so if anyone sees a bird, just call it out." He went on to emphasize the importance of listening to birds by describing a bird walk he had done with another walk leader: "It was so successful because he was really good with his ears, and I'm really good with my eyes, so between the two of us, we were able to observe a lot of birds." Bill and his wife engaged in this kind of teamwork when, on one walk, we heard a Cardinal. Bill pointed out the sound and said, "It sounds like a car alarm." Bill headed up to the front of the group explaining the sound, and Sharon stayed at the back of the group. She called up to Bill, "There's a Cardinal!" and pointed at the bird. Bill laughed and said, "I heard it, and she saw it!"

People often hear birds singing. Birdsong announces springtime, or almost any lovely day. To birders, birdsong means much more. Sounds indicate birds' presence or absence, as Catherine noted when leading a walk: "It's so quiet. I'm not sure if we'll see much, because it's so quiet." Different types of sounds also mean different things to birds, such as singing to attract a mate or chip notes to warn other birds of danger. Before deciphering bird sounds, walk participants must learn to discern bird sounds from other background noises. This process of separating the primary "figure" from the "background" can be both visual and auditory.[5] People often tune out background sounds, and there's even an entire genre of background, or elevator, music.[6] Therefore, people must engage in "mental focusing" to give their visual and auditory attention to a particular item.[7]

Birders must pay special attention to bird sounds, which people typically ignore. In our interview, Miranda, the executive director of a local Audubon chapter, explained how birders develop their attentional skills. She said that even though people appreciate birds, they don't have a "general awareness" of birds, because it's not "socially acceptable, or not trendy" to be a birder. She contrasted the awareness people may have of other elements of the outdoors with their comparative lack of awareness about birds: "People will go outside, and they feel the sun on their face, and they feel the wind blowing in their hair, but they don't necessarily hear—they're not listening. They hear the birds, but they're not listening to the birds. They're not really appreciating the birdsong, or trying to figure out what bird that they're hearing. They may see birds, but they don't really appreciate the different species that they're seeing. So, it's kind of this background that I worry will slowly disappear. Like Rachel Carson, a *Silent Spring*. People aren't going to realize it's gone until it's too late." Miranda referenced Rachel Carson's book *Silent Spring*, which warned of the effects of DDT and other pesticides on birds and other wildlife.[8] If we did not pay attention now, Carson warned, we would notice only when springtime arrives and no birds sing. Miranda said that while people may hear birds, they aren't listening to birds—they aren't paying attention. This active attention helps birders find and identify birds, and as I demonstrate in the next chapter, it helps birders cultivate an ecological sense of being in the world by developing a naturalist gaze.

Active attention helps people appreciate even more than they thought was out there, as Cynthia noted from leading walks with high school students. Cynthia works as the youth education manager for a local Audubon chapter, and she described herself as a "meanderer" when she takes nature walks or hikes. She tries to instill attentiveness in her students as well: "I'm always touching things, or smelling things, but I will really make an effort to listen and I'll actually stop. And I make an effort to have, especially our high school students, to stop because we'll be on a bird walk, and they'll be complaining, 'We're not seeing anything!' and it's like, 'They know we're here, so if we just stop for a few seconds and be quiet,' and then they hear the woodpeckers. And they start to appreciate that." Cynthia says that learning to pay attention to each rustle, even if it's a squirrel, might help you notice a woodpecker in the forest, and that before she became a birder, she said, "there are times when I just wouldn't notice that." Miranda and Cynthia tried to remind themselves—and others—to pay attention to the background noises that they might typically ignore. Such attention opens up a new world, as Zerubavel says, "What we explicitly notice is proportionally smaller than what we implicitly ignore."[9]

Walk leaders know people go on guided bird walks to see as many birds as possible, and thus they teach this attention to participants. Walk leaders set the stage for active listening by reminding participants to stay quiet and listen, as Diane did at the beginning of each walk she led: "This is a large group, which is good, because we'll have more eyes, but we have to keep our voices down because we'll also be birding by ear."

Sometimes participants thought that only hearing a bird did not "count" as an observation. For example, Justin explained why the group wasn't seeing many birds on a particular walk, and he referenced a breeding bird count he had just participated in in New York City. "They're staying home, tending their nests and their young, so we can only hear them. You have to do most of your breeding bird surveys by ear; that's how you get 90 percent of the breeding birds," Justin said. "Is that legitimate?" one participant asked, and Justin replied yes. The participant worried about the veracity of the count if the observer only heard and did not also see the bird.

More often, participants simply weren't satisfied by only hearing a bird. They wanted to see the birds as well. At the beginning of another walk, Justin said, "We are going to walk around and try to see as many birds as possible, and more importantly, hear them." One of the women on the walk said, "Or, more importantly, see them! I'm not going to be a happy camper if I don't see them." Later in the walk, Justin pointed out some bird calls and said, "Sorry we aren't seeing a lot of birds, but it's sure nice to hear them!" Walk leaders tried to instill an appreciation of hearing birds, rather than being satisfied only with seeing birds.

However, appreciating birdsong rather than seeing a bird was almost a necessity for certain species of birds, such as Warbling Vireos. As Justin explained, "They are very cute birds, but they hide a lot, so it's hard to see them. But you can hear them." On that walk with Justin, we only heard the Warbling Vireo, and for the next two walks I attended, each group had the same result: we heard this bird, but we did not see him. But on the next walk, the group finally saw one, even though some participants didn't think it would be possible. As the group walked down the path, one of the participants heard a Warbling Vireo, and she announced his presence to the group. A man on the walk said that he really wanted to see one, to which Marie, a longtime birder with over twenty years of experience, queried, "You're going to try to *see* a Warbling Vireo?"

How did these walk leaders, and so many walk participants, even know what type of bird they heard from the bird's call? Birding by ear represents a more advanced method of birding, where birders can identify the exact bird species from his or her call. This knowledge can lead to a special kind of frustration for birders when they hear a Red-tailed Hawk's strong cry dubbed over a Bald Eagle's weak, high-pitched whistle in movies and television shows.[10] Or, this knowledge can help solve crimes, as when forensic analysts ask ornithologists and birders to identify birds in the background of telephone calls or other recordings.[11] Sometimes, birding can be a completely auditory experience, in the case of blind birders. Birding is promoted as a pastime for individuals who are blind, since each bird call is unique. Blind birders lead birding by ear workshops for other individuals who are blind, and sometimes they form teams to participate in birding competitions.[12] On one walk, when Bill was encouraging the group to practice birding by ear, he said that a friend of his has a daughter who is blind, and so Bill suggested she get into birding. Bill's friend asked why that would be a good hobby for someone who is blind, and Bill said it was because of the birds' calls. She would be able to tell the different types of birds because of their calls—she could listen to them and tell the exact type of bird.

For birders who are fully sighted, learning to bird by ear gave them a deeper appreciation for birds and for all of the sounds of nature, as Trish, a retired microbiology professor, explained in our interview: "It's given me a way to appreciate it more. Like I said, it's really helped train my eyes and ears, so that I see and hear more. And I'm sure I still need to do better and better at that. But I think that it's helped tremendously in that way, especially in the last ten years, since I've really tried to learn to bird by ear more. I've realized what I had been missing before, not even picked up on, not only birdsong but all kinds of other noises that happen when you're out in nature. So, I think in that way, birding has definitely affected my relationship with the environment."

Birding by ear is still a skill that takes practice, and several Audubon chapters and other birding clubs offer workshops on birding by ear. I attended one such workshop led by Diane, who began by saying that birding by ear is a "great way to tap into the rhythms of the earth." Further, "birding by ear allows us to bird in 360 degrees, but birding by eye only allows us to see what is in front of us." Diane explained various reasons why birds sing, such as creating a "sound fence" to defend their territory or singing to impress others. Birdsong can show a bird's health, and birds have multiple songs, which can vary geographically. Understanding birdsong is especially important for spring breeding season, when birds continually sing to impress a potential mate. Listening to birds helps birders learn to pay attention to things we typically ignore. This careful attentiveness composes part of the naturalist gaze.

HOW TO FIND A BIRD TO SEE

Paying attention to all of the sights and sounds of nature helps birders find birds to observe. The first step in watching a bird is to find the bird, which is often difficult if the bird is in a leafy tree or bush. Several walk leaders encourage participants to use a clock face to point out birds' location. Bill uses a laser pointer to point near (but not on!) birds in bushes and in trees. Every time Bill uses the laser, participants note how useful it is for finding birds, and Bill replies, "That's my job, to get you to see birds." For example, on a walk with Bill as the leader, someone in the group pointed out a Goldfinch on a tree. Bill located the bird and told the group the bird was at twelve o'clock on the tree, but several people still had a hard time finding the bird. Bill got out his laser pointer and used it to point out the bird, allowing everyone to find it. Someone said, "It was hard to see at first, because the Goldfinch was the same color as the leaves." Bill said he uses the laser for "novices," and from my observations, he uses it only as a last resort, when several people in the group have difficulty finding a bird.

More often, walk leaders teach birders how to find birds by giving tips on how to use binoculars. On my second-ever bird walk, and my first walk with my new binoculars, the group stopped in the middle of the trail while everyone murmured admiration for a Palm Warbler. Everyone was enjoying this bird, and I felt frustrated because I couldn't find him in my binoculars. "Did you find him?" a woman next to me asked. "I can see him when he's flying from branch to branch, but I can't find him with my binoculars," I told her. "That's a good skill to have, to be able to see a bird with your eyes, and then keep your focus on the bird, and bring your binoculars up to your eyes," she explained. I practiced a few times and finally got it. Binocular skills take practice, and many walk leaders encourage practicing binocular skills at the beginning of the walk, as Justin did before a walk: "Let's practice on that rock. Keep your eyes on the rock, and just bring the binoculars up to your eyes." In our interview, Bill told me even more about binocular use:

> Make sure your binoculars are clean, make sure you adjust your binoculars. I constantly scan the trees, not that I'm looking for birds necessarily, but I'm looking to say, well, if the bird was over there, I already have my eyes pinned there. And you get practice. Look up at the tree, right? You see the bird moving? Just bring the binoculars up to your eyes. And then when you don't see it, bring your binoculars down. Look for the movement. It's over there. Again, practice. Practice with your binoculars. What kind of binoculars do you have? If they're too big, they're not that good. You want something like 8 × 40. You want something where the big number divided by the small number is as close to five as you can get. Do you wear glasses? You roll the cups down. I said to this one woman, "Do you wear glasses to bird?" She said yeah. I said, "Well, roll the cups down and see if that's better." "Oh, it's great!" Because it brings the binoculars closer to your eyes. Because those cups are down, but if you don't use glasses, you keep the cups up. So all of those things.

My inability to find a bird in my binoculars was due to my inexperience using binoculars, but other times it was due to getting accustomed to a new pair of binoculars. One walk participant admired the new binoculars of another participant, who said she "treated herself" with her L.L. Bean points to purchase the new and expensive binoculars. She explained, "I sacrificed depth for strength, and it takes a lot of practice to use them. I'm not very good at being able to find something quickly in them. I didn't get to see the Red-bellied Woodpecker in them, but I saw it with my bare eyes."

Another skill birders learn is how to discern birds from the background. As one raptor expert said, when giving a presentation to a local Audubon chapter, "I guarantee you've walked right past a Great Horned Owl and simply didn't notice it because you weren't looking for it. They're designed to blend in to the trees." How do birders discern a bird in the first place, especially if they are "designed to blend in" to their surroundings? Typically, walk leaders tell newcomers to pay attention to any movement with your bare eyes, and then bring your binoculars up once you've found the bird. This is difficult to do if you're the only person who has found the bird. On one walk, we were on the lookout for a Barred Owl who had been seen in the pine forest of the park. Phyllis, the walk leader, suggested we split up and fan out to try to find him. "Look out for white droppings near the bottom of the tree," Warren explained, as a sign that the owl had been perched there. All of a sudden, Richard exclaimed, "I got him!" I was right next to Richard and was able to just see the owl flying to the next patch of forest, but I couldn't see where he landed. Everyone came over to where Richard stood, still watching the owl through his binoculars. "Where is he?" everyone asked Richard. "I'm on him, but I can't take down my glasses to point him out, or I'll never find him again," he responded. Richard tried to describe the particular tree where the owl was perched, but no one could find him. "You'll never see him with your bare eyes unless he flies again," Richard said. "Do you want me to call him and see if I can flush him out?" he asked the group, who said yes. Richard did an impressive imitation of the "Who cooks for you?" call of the Barred Owl, but the owl flew off in the opposite direction, and not toward us.

Richard's intense focus on the bird through his binoculars prevented him from being able to describe the bird's location to us, among dozens and dozens of other trees. Moreover, the bird's feathers and lack of movement gave him perfect background matching, allowing him to blend in to the trees and remain unobserved.[13] While Zerubavel notes that people purposefully engage in background matching when they want to remain inconspicuous, he uses birds' natural camouflage as the basis for his human examples.[14] Birders, too, joked about birds' "camouflage," such as the difficulty of finding warblers in trees where the leaves are the same size and color as the birds. While birds' plumage purposefully serves as camouflage, it can also inadvertently foil birders when birds' plumage takes on an unexpected color. While leading a walk, Luis pointed out a warbler in the tree canopy and said, "He's not doing us any favors—he's in fall plumage and is molting." Fall plumage is more muted than springtime breeding plumage, and molting means a bird is losing his feathers and growing new ones. In other words, this bird had few feathers, and those that existed hardly looked typical. Luis found the bird again and explained she was a female Pine Warbler, and females have even more muted coloration than males. Someone on the walk then asked, "How can you tell what it is, if it's molting and in fall plumage?" Another person responded, "Thirty years of experience," and everyone laughed. The trick of using a clock face for location, tips for how to use binoculars, and attentional skills to distinguish birds from their background all help birders find birds to see.

IDENTIFYING CONTEXT WITH THE NATURALIST GAZE

Seeing and hearing birds are not the only activities for birders. Birding also involves identifying each species of bird one sees or hears.[15] Birders typically record their species lists for their own records or submit them to various citizen science projects.[16] More than binocular skills or deciphering camouflage, being able to understand all of the aspects of a particular habitat, combined with years of practice, created the best conditions to spot and identify birds. On another walk with Luis, we stopped at a patch of jewel weed, with its bright orange flowers, to try to spot some hummingbirds. We stood there for quite some time but did not observe any movement. Suddenly, a robin-sized bird flew by, so close that everyone on the walk exclaimed "Whoa!"

"That was a Catbird," Luis said. I wondered how he knew that from such a brief glimpse, and then someone else asked that very question.

"How did you know that? I just saw a dark shadow and couldn't tell what it was!" she said.

"Part of it is habitat, part of it is its flared tail, and part of it is that you can hear Catbirds in the area. Even if it were a Cardinal, you'd be able to see some red. It wouldn't be a Blackbird because they don't like rushes. There are only a few other birds this size among the passerines we encounter here," Luis explained.

Luis pieced together several pieces of information that as a birder, in this particular attentional community, he had the skills to know what to pay attention to. He knew to pay attention to habitat, color, size, and tail shape. In this way, Luis has developed skills such that he could engage in automatic spatial cognition and immediately identify a bird, in what seemed an impossible feat to the rest of us.[17] This ability to place birds in context, and use context clues as an aid in identification, also makes up part of the naturalist gaze.

Earlier I described how birders find birds by sound and by sight. This section explains how birders learn to identify birds. The first step in identification is learning to accurately describe birds, using specific language. For example, on my very first guided bird walk, the leader, Bill, asked us to call out any birds that we saw. I naively pointed out a "seagull," and Bill jokingly chastised me: "*Sea*gull? We don't say 'seagull.' Would you say 'bay gull' if you were on the bay? Instead, we say 'Ring-billed Gull' or 'Black-backed Gull.'" I'm still not quite sure if Bill was making a "bagel / bay gull" joke, but his response demonstrated the importance of specificity in language.

Novices use unspecific language to describe birds. On a different walk, a new participant pointed out a bird flying. Someone asked what type of bird it was, and she said, "I don't know! It's small!" and everyone laughed. She said, "It's just like if someone asked me about a car—'It's a red car!'" Another time, when I described a bird as brown, the walk leader asked me if that meant mahogany or cinnamon or a different shade of brown. More experienced birders, including the walk leaders, held themselves to the same standards as they applied to us, expecting specificity in their own language. When a novice pointed at a bird and asked, "What is that little bird?" Diane, the walk leader, responded, "Try to use more specific terms. I have to break myself of that habit as well." Or on a walk with Luis, he said, "Fall migration started a few weeks ago—excuse me, *passerine* migration. So we should be seeing lots of warblers and other passerines flying south for the winter."

Why is specificity important? One reason is because 9,000–10,000 species of birds exist in the world, and many birders have thousands of species in their repertoire of birds. For example, throughout 2015, the National Audubon Society covered the story of Noah Strycker on his "big year" quest.[18] After traveling to all seven continents, Noah saw 6,042 birds, setting a new world record for a big year. With this in mind, Luis's encyclopedic knowledge about subspecies of Canada Geese becomes more understandable. As we passed by more than fifty Canada Geese in a field while on a walk, someone asked Luis, the walk leader, why so many geese were here. Luis said, "They're going to eat nonstop until their fall migration, towards the end of November." Luis then began listing a dozen different subspecies of Canada Geese that we might see in the park over the fall.

Another reason for specificity is because many birders participate in citizen science projects, recording and submitting their sightings to organizations such as the Cornell Lab of Ornithology through eBird, or the National Audubon Society through the Christmas Bird Count. Birders want to be as accurate as possible in these lists, because the purpose of these citizen science projects is to help birds. It does not help birds to lie about which species they saw, or how many of them they saw; thus, birders maintain integrity and accuracy in their lists, to the greatest extent possible.

A third reason for maintaining specificity is simply the personal satisfaction of getting it right, of solving the identification puzzle. In this way, birders are also intrinsically motivated to identify precisely which bird they saw or heard. Figuring out the species is part of how birders fully appreciate their finds. For example, at the annual meeting of one local Audubon chapter, a couple described some birds they saw at a local lake that they could not identify. They said they went home and looked up the birds because they couldn't identify them, and Tom replied, "Isn't that the most fun? To go home and look it up and identify what you saw?" Everyone heartily agreed. Unlike the passive observation (or sometimes willful ignorance) of animals in zoos, birders find fun in observing birds and then learning more about them through research after a walk.[19]

Birders employ social media to help them identify birds when they've exhausted their own resources. The Facebook Bird ID Group of the World helps birders make accurate identifications. Birders post a photo of the bird, along with the date and location, and other group members help the original poster identify the bird. Just as walk leaders try to develop participants' birding skills, the group rules ask that commenters not blurt out the identification, but rather provide hints so that the original poster can figure out the identification on his or her own: "Please do not comment by simply stating the name of an ID. Commenting 'Mallard' doesn't help posters and others in the group understand how to identify the bird. Instead, we ask that you either help guide the poster to the ID with constructive comments about field marks or thoroughly explain your identification. Please provide more details than just saying 'check warblers.'"[20]

Part of why birders can experience intrinsic satisfaction from bird identification is because birds are knowable. It's entirely possible to identify every bird that a birder sees. However, it takes skill and experience to be able to do so. Walk leaders, social media commenters, and, of course, field guides assist birders in completing their identifications. Field guides represent an external, institutionalized classification system for birders, providing criteria for birders to categorize birds.[21] Identifying birds on the basis of a scientific classification system can begin for birders as "deliberative cognition," but birders can become so fluent in the categories and classifications that identification becomes based on more automatic schemata.[22] As I show later in the chapter, expert birders can automatically identify birds, and they then explain their classification choices to other birders—to novices wondering how they came up with that identification, or to other experts seeking to challenge or verify the identification.

As an external classification system, scientifically based species categories are fairly top down, imposed on birders with little opportunity to provide feedback on the categories.[23] The binomial nomenclature system, which includes genus and species, is now commonly known to birders, even though birders and ornithologists typically use birds' common names in practice (e.g., Bald Eagle instead of *Haliaeetus leucocephalus*). In the late nineteenth century, the adoption of this system was one of the primary methods the burgeoning professional field of ornithology used to separate professionals from amateurs.[24] Professionals named birds; amateurs used those names. Birders sometimes express frustration at the categorizations, lamenting name changes. For example, on a walk, someone spotted an Eastern Towhee. Another birder said, "I learned that bird as a Rufous Towhee, and then the name got changed to Eastern Towhee. I learned it because my old field guide had the old name in it." The walk leader chimed in, "That reminds me of a walk when someone asked about an 'Oldsquaw Duck,' and I had to tell him, 'There are no Oldsquaw Ducks around here, only Long-Tailed Ducks,'" referring to the official ornithological name change.[25]

Field guides can be daunting, especially to novice birders. In field guides, such as the *Peterson Field Guide to Birds of Eastern and Central North America*, birds are presented in groups, such as "Geese, Swans, and Ducks," "Shorebirds," and "Cuckoos," but also "Rail-Like Birds," "Alcids," and "Tanagers and Miscellaneous Passerines."[26] In this way, field guides and birders engage in mental "lumping" together of certain birds, and mental "splitting" apart of other birds, in order to differentiate among different types of birds, and then among species.[27] What if you don't know what type of bird you're looking at in the first place? What's a "Rail-Like Bird"? How do you even know where to begin? Dawn, a PhD student in environmental studies and the director of conservation for a local Audubon chapter, explained to me in our interview some of how she learned to make identifications:

> Once you have to identify them, you have to pick up subtle differences between species, and so you start to really observe their behaviors and their body shape and their, how they carry themselves, and what their beak shapes are like, and why their beak shape is like that, because of food, because of what they eat. And so, I think if you're really, truly observing anything, your knowledge about them increases dramatically. Not only because you're interested, but also to identify it, you pick up all these subtle cues and differences between them, and then those differences, there's a curiosity of why they have that difference, and then you learn about what that difference is.

Dawn described a number of elements that go into the initial steps of identification, and she said she shares some of these "fun facts" when she leads bird walks, such as the "cool facts about unique adaptations" that different species have. Novice birders learn these tips through participating in bird walks, as I describe in more detail later in the chapter, and field guides also provide some guidance for beginners. The *Peterson* field guide suggests that birders pay attention to a bird's size, shape, wing shape, bill shape, tail shape, behavior, whether it climbs trees, how it flies, whether it swims or wades, its field marks, tail pattern, rump patch, eye stripes and eye-ring, wing bars, and wing pattern. That's a lot to pay attention to. Bill said he prefers the *Peterson* field guides because they help narrow down the range of exactly what to pay attention to. On a walk, Bill explained that the *Peterson* field guides distinguish themselves from other field guides because they have arrows to show birders the exact spot to look for in order to identify the bird. Or, as the *Peterson* field guide puts it, "The arrows point to outstanding field marks."[28] This means birders don't really have to pay attention to all of those aspects, but instead only the distinctive field marks that identify that particular species. Bill explained to the group that Peterson had trademarked the arrows, and so no other field guides could use such diagnostic arrows. In our interview, Bill explained some other tips for identification:

> That's why I like *Peterson's Field Guide* again, because he has a lot of those silhouettes. And the habitat, I had a nuthatch walking down a tree, he has a woodpecker walking up the tree right next to it. So you see the silhouettes, you say,

> "What are they doing?" I always say, "Where is it? On the ground? Up in the air? Is it walking up the tree? Down the tree?" That gives me a hint of what's happening, if I don't see the bird. Then I start saying, "What color is it?" But if the person says it looked like a sparrow, or any particular bird, yes, get the pigeon down, the sparrow, and the robin. Because you can say was it in between a robin and a pigeon size? Between a robin and a sparrow size?

Bill reiterated many of the identification tips from the *Peterson* field guide, but he also provided another important method that beginning birders learn to use for initial bird identification—comparison with other, common birds. Bill also suggested comparing a bird with a sparrow, robin, pigeon, crow, or hawk to describe a bird's size. Common birds serve as a vehicle for practicing proper bird identification.[29] If birders can take the time to observe the behavior and plumage of common birds, they can translate those observational skills to other, less common birds, and they hone their overall birding skills. Just as individuals understand different elements of a cultural system by using common attributes and relations to others, so, too, do birders understand different elements of a zoological system.[30] Birders learn to identify birds' common attributes and how they differ from other, similar birds.

The Merlin Bird ID smartphone app from the Cornell Lab of Ornithology uses this technique as well. The app is designed for beginning birders. After confirming the date and location of the sighting, the app first asks birders to estimate the size of the bird, using silhouettes of other, common birds as a comparison. The size silhouettes range from a sparrow, to a robin, to a crow, to a goose, with choices for sizes in between. Then, the app asks for up to three colors on the bird, and it finishes by asking, "Was the bird . . . ?" with responses including "eating at a feeder," "swimming or wading," "on a fence or wire," "soaring or flying," and so on. After taking all of this information, the app generates a list of potential birds. Birders can look at photos and read more about each bird, and if the correct bird is there, the birder can click "Yes, this is my bird" to confirm the identification and help the app hone its accuracy.

Birders learn and develop these identification skills on bird walks. While field guides and identification apps may help, birders primarily learn to identify and differentiate birds while on guided bird walks. Many different species look similar, so birders often share tips on how to tell the difference between similar-looking species, such as crows and ravens. On one walk, Tom saw a nest and said he thought he saw a crow flying by earlier, so maybe it was a crow's nest. Catherine said she only recently learned how to tell a crow apart from a raven, and the key was to look for the rounded tail.

Understanding how to differentiate between species, or even how to see the markings of specific birds, represents an important skill for birders to develop. Walk leaders emphasized the importance of understanding markings and identification for common birds just as often as they did the more exciting and rare birds.[31] One of the most common birds in the New York metropolitan area is the American Robin. Still, walk leaders take the time to observe American Robins and point out how to identify them. This is a bird that almost no one needs help identifying, but walk leaders explain very specific aspects of their plumage and behavior so participants better understand them. On one walk, a woman stopped, looked at a bird, and asked the group to come over. She said, "Do you see that bird with the red breast?" Luis said, "Yes, that's a female robin." The woman said, "So a male robin would have brighter colors?" and Luis replied affirmatively.

Walk leaders explained their bird identifications to walk participants as a way of teaching participants how to identify birds. Walk participants also explained their identifications to the rest of the group. When anyone spots or identifies a bird, they explain why they made that identification. On one walk, someone pointed out a gull and exclaimed, "There's a seagull!" to which Diane, the walk leader, quickly corrected, "Don't say that!" at the same time that he was correcting himself and saying, "Or, a gull." Diane asked, "Which type of gull do you think it is?" and the man said, "A Black-backed Gull." Diane then prodded, "What made you think that?" and the man replied, "The wings, and how he's soaring." Diane was not challenging the man's identification—she was inquiring about the characteristics he used to make his identification.

Birders become so accustomed to explaining their identifications to others that they automatically respond in that manner, even when that isn't the question asked of them. As we approached a lake on an early morning walk, with the sun rising, the bright light bounced off the water, making it difficult to see. Everyone was looking at a tree where a man had pointed out Yellow-rumped Warblers. A woman asked, "How can you tell?" and the man said, "Because of the yellow back, and stripes." The woman interjected, "No, I mean, how can you see, with all of this backlighting?" She was simply asking how in the world he managed to see the bird at all, with such bright light around, and was not asking him to explain his identification.

Birds are knowable, but bird identifications are not foolproof. Birds do not sit still and display all of their distinguishing characteristics for onlookers. Even if they do, a level of uncertainty always remains in identifying a bird. Birders use all of the tools available to them to identify birds—conferring with other people, writing down specific descriptions, looking in field guides, and so on. People hold a strong aversion to ambiguity, and so we always try to place people or objects into categories.[32] Birders have to learn to live with a certain level of uncertainty in their identifications.

Because birds are knowable, but our knowledge is fallible, bird walks often include debates over identification. Birders don't always resolve ambiguity, even when multiple people participate in the identification attempt. On one walk, people kept hearing a Grosbeak, but we never saw it. We heard the Grosbeak again, and someone saw a bird in a tree that they thought might be the Grosbeak. Everyone kept offering ideas as to how to discern what bird it was. They got out the spotting scope for a better look. They asked more people to look, to get multiple people trying to identify it. They looked through a field guide. They took a photo of the bird and passed the camera around. They got out their smartphones and looked up identification facts about the bird. No one could definitively say what type of bird it was. They kept talking about the bird's stripes and the white around his eyes, but no one could tell. After a while, one man said, "Is the verdict in? Is it a bird?" and everyone laughed.

Ambiguously marked birds often leave birders perplexed, left with a problem to be solved by further research. Or, birders must simply learn to live with the ambiguity and uncertainty of not being able to identify every single bird that they see. On a walk with Bill, Janice stopped and asked Bill for help identifying a flycatcher. We all looked at the flycatcher for a while, but no one could tell exactly which kind of flycatcher it was. "I think it's a Least," Janice said, meaning a Least Flycatcher. "There are five different types of Leasts, which one do you mean?" Bill responded. "We'd have to hear the song. I think it's a Pee Wee, but you'd have to hear the bird." To accurately identify a bird, Bill explained, "you need the visual, the song, and the habitat. It's odd that the bird isn't singing. You have to hear it—even experts need to hear it. It's the only way." We all kept looking and waiting and looking at Bill's field guide, and then Bill said, "Sometimes you have to be comfortable saying, 'I don't know.'"

Bill's suggestion was easier said than done. Getting to "I don't know" first meant having a debate over the identification or trying to lure out a bird using playback, with the hope of getting a better visual and then being able to make the identification. Ultimately, many potential identifications end in "I don't know." In this way, birders learn to develop what Zerubavel calls "the flexible mind," or the ability to see the world as sometimes quite clear-cut and other times as undifferentiated.[33] Even though all birds are knowable, they are not always identifiable, and birders learn to live with that ambiguity. They learn that sometimes they can't identify a bird, they can't add that bird to the species list, and they simply won't get the bird on that day. These context clues that help birders learn to identify a bird, and the ability to leave a bird alone without getting the identification, also forms part of the naturalist gaze.

BIRDING AS FLOW

Walk leaders enjoy sharing their love for and knowledge of birds, but they also seek to develop self-reliance in the newer birders attending the walks. In contrast to the earlier stories of teamwork in finding and identifying birds, walk leaders instruct newer birders how to develop as birders more generally, including not relying on field guides or other people to help them find and identify birds. They even discourage birders from being too excited about seeing birds in their binoculars. For example, before beginning one walk, Justin, the guide, instructed the group on what he called the four stages to being a birder: "At first, you are learning how to use your binoculars, and you are so excited to see birds. Then, you rely less on your binoculars and you learn more about bird calls." He demonstrated this by listening for several different bird calls and identifying each of the birds. "After that, you can start to tell birds by their silhouette." Justin then demonstrated this skill by quickly finding different birds by shape and pointing them out to everyone. "And then finally, you understand all of this, and you can go for a nice walk in the woods and it becomes much more Zen. Then, you don't need to carry your binoculars all the time. You can just take them out when you need them." Justin demonstrated this final stage by pulling his small, unobtrusive binoculars out of his pocket. One of the attendees said, "I've been birding for a long time, and

I love my binoculars." She held up her large binoculars that were strapped to her chest. "I don't want to give them up!" Justin laughed and said, "Don't worry, I won't make anyone give up their binoculars."

In this introduction to the walk, Justin demonstrated several birding skills, such as birding by ear or by silhouette. He called the excitement of visually seeing birds one of the beginning stages of birding, which eventually, as one progresses in the stages, will culminate in a lovely walk in the woods where the observer takes note of birds through multiple senses, and not always or only through the lenses of binoculars. Although a person holding binoculars to the sky remains a stereotypical image of a birder, Justin presented birding as a Zen-like walk through the woods, where the astute birder can experience the joy of birding in multifaceted ways.

Another stereotypical image of a birder is that of a person leafing through a field guide, trying to find the bird they just observed. However, during a walk with Luis as the guide, he gently admonished participants for relying so much on their field guides, and he encouraged them to develop their own powers of observation. As a couple of people desperately tried to find a particular bird in their printed field guide, Luis said: "I don't mean this to sound critical, but I don't recommend using field guides in the field. First, you want to see the bird. You have to learn how to see the bird, and then you can describe it, and I can help decode it. Field guides don't matter if you don't actually get to see the bird. The best skill is to learn how to get on the bird. That takes time, especially for quick birds like warblers. They have such a high metabolism—not as high as hummingbirds—that they are always moving and always eating. To find insectivores like warblers, you have to find where the sun hits. That warms up the insects." Why would Luis discourage participants from using field guides when field guides and binoculars are the two most distinguishing characteristics of a birder? Luis was not trying to get the participants to solely rely on him. As he later said, "The hardest thing is to go birding with a more experienced birder and still be an active observer. It's too easy to just sit back and let someone point everything out for you." Luis equally did not want participants to solely rely on him. Instead, he wanted them to develop their active observation skills so they could improve as birders. He said that he has a thirteen-year-old daughter, and when she was born he went on a seven-year hiatus from birding. "Of course I continued to notice birds, but I didn't go out as much as I used to," he said. When he returned to birding, he said he did not bring his field guide. He only wrote down descriptions of birds and drew sketches, and then studied at home to figure out which birds they were. He said he learned more from doing that than he did in ten to fifteen years of studying field guides.

Even though Justin and Luis were serving as leaders on guided birding walks, they were trying to develop the participants as birders by creating active learners who no longer needed them as guides, or even needed binoculars and field guides while in the field. They sought for participants to be able to describe birds, to learn how to hear and distinguish birds, and to be active observers through all of their senses. This active observation and learning contrasts with the passive learning about animals at zoos and aquariums, where wild animals in captivity are presented to audiences in unnatural settings.

The culmination of mastering these skills and becoming a birder means that birding becomes what the positive psychologist Mihaly Csikszentmihalyi calls an autotelic experience, or "flow."[34] Flow activities are done for their own enjoyment and not necessarily as an end in themselves. People describe flow experiences as when they get lost in the activity, concentrating so intently on it that all other concerns drift away. We can experience flow at work or at leisure, but some activities lend themselves to producing flow more than others. Earlier, when Luis and Justin depicted how the ultimate experience of birding meant being one with nature, paying close attention to nature, and not getting caught up in the use of binoculars or field guides, they were describing birding as flow. As they noted, flow comes at a more advanced stage of becoming a birder, and birders must master certain skills first. Flow results from structured activities more often than by chance.[35] Moreover, flow activities are designed to create such experiences because they allow participants to achieve growth and discovery. Birding presents an excellent example of a flow experience, as it incorporates several elements that lead to flow.

As noted in the sections about learning how to bird, birding takes skill and presents challenges. Flow activities should have a good ratio of challenge to skills—they should be challenging, but not too challenging.[36] Birding also presents clear goals to participants and provides them with opportunities for immediate feedback.[37] Birders go on walks with the goals of seeing, identifying, and recording different bird species, and they experience immediate feedback of the satisfaction when they successfully achieve those goals and when they submit their species lists for various citizen science projects. Even just seeing an exciting bird, or the "goal bird" for the day, can provide such positive feedback. The challenge of birding also provides benefits beyond experiencing flow: researchers have found that birding reduces stress, and the more species a birder knows, the more they experience such enjoyment and stress reduction.[38]

Flow activities also allow people to completely lose themselves in an activity. People so intensely concentrate their attention on the activity that they can even lose a sense of time.[39] This does not mean that flow experiences are easy or effortless. Rather, they can require intense physical or mental exertion. They only appear to be effortless. While birding does not require such physical feats as rock climbing (Csikszentmihalyi's example), birding can be physically uncomfortable, such as on early, cold mornings, as Ellie, a birder who recently retired from book production, described to me in our interview. Because she was so caught up in the joy of birding, she was able to forget all of that:

> We went up to, a couple years ago, up to the Black Dirt Country on a day in maybe January, maybe February, when the day started out really cold. And we got up there before dawn, in order to see Long-eared Owls taking off in this big field. We could barely see them. We were freezing to death. Some people saw more than I did. I don't know why I was having trouble seeing them, but I didn't really see them. But then the sun came up, and it turned into this beautiful, sunny, but cold, day. And we went to various spots in the area, and at some point we were walking through, between fields, corn fields, with irrigation ditches in between, like lots of, I mean the black dirt is really peat. It was still freezing, well below freezing outside, and one of the irrigation ditches was right next to us. And frozen over. And the blue sky reflected off of it and right down ahead of us to the right was a Great Blue Heron, somewhat carefully stepping on the ice. And it was just so beautiful. And then there's a little steam rising up from the peat, from the black dirt in this cold, cold air.

Ellie remembers the steam from the peat, and she described catching a glimpse of a mink. Even though she didn't get to see as many Long-eared Owls as she was hoping, she remembers the beauty of that day, in the moment. She described birding as a "form of meditation," or "just being in the here and now." She didn't dwell on the extreme cold she was in for hours on end; she lost herself in that moment. She said, "You don't remember being cold and miserable." One birder, writing a newspaper column, told a story about being so caught up in birding in the Everglades that he didn't realize he was about to walk into an alligator.[40]

Ellie, along with most of the birders I interviewed, talked about birding as a meditative activity. Flow activities exemplify a paradox of control, meaning that we have a sense of control over the activity that we don't typically feel in our everyday lives.[41] Many birders talked about birding as a way of escaping their everyday stresses, as Toby did: "It's relaxing. It puts you in the moment, and you forget about everything. It's always been that way. For me, going, psychologically, going all the way back. You have a hard day, or whatever, go out bird watching, out in nature, it just refreshes you. And I'm forty-seven now, I started this when I was, really more seriously when I was ten, so my whole life, I've always, any hard times, I've always turned to a hobby like birds. And it's very relaxing. It just helps clear your mind, get better perspective on things. It's very therapeutic." Birding differentiates itself from "forest therapy" or "forest bathing," a practice started in Japan, in that birding requires skill and concentration.[42] Forest therapy encourages people to sit still in a forest, with "no need to think or analyze."[43] While people may enjoy similar de-stressing benefits from both birding and forest therapy, only birding provides such a flow experience. Birding requires concentration and involvement, and this intense concentration allows participants to forget about their worries or stresses.[44] Walking through the woods, listening intently for any sounds, watching carefully for any movement, and then focusing on the bird and deciphering the bird species all create a flow experience.

At the National Audubon Convention, I interviewed Allen, a wildlife biologist with the National Park Service. Because birding creates a flow experience, even if he doesn't see a bird, Allen said, he still enjoys the meditative aspect of birding: "There's been studies about how nature affects people, and how it recharges you and clears your mind and values along that line. So I think it's, to me it is very relaxing. I'm not a religious person by any means, I don't mean to call birding my religion, but it's kind of that way, where it's, it just gives you a way to look at things and relaxes you. It's more like a meditation kind of thing, where it clears your mind and makes you feel better. And then, especially if something is frustrating at work or at home, it's just a way to reset and refresh." Allen said he's still content if he goes out birding and has "a nice walk and I don't see a single bird," because, as he explained, birding was part of a larger experience of "being out in the trees." While seeing birds is enjoyable, the practice of going out into nature and actively paying attention to nature helps birders lose themselves in the experience. As Kristina told me, "That, to me, is the most important thing about birding, is taking a moment of time and really just being present at that moment." Birding presents a prime opportunity to enjoy a flow experience, to lose oneself in the moment, by paying close attention to the sights and sounds of nature and using one's skills to find and identify birds.

This chapter demonstrated how someone becomes a birder. Through practice and teamwork in guided birding walks, birders learn how to hear and see birds, how to find the birds they wish to see, how to accurately observe and identify birds, and how to develop active observation skills. The ultimate goal or stage for birders was a flow experience, to be able to enjoy a "Zen" walk in the woods and appreciate the birds that one hears and perhaps sees, without relying on field guides, other people, or even binoculars. The guided birding walks help teach birders the skills they will need to be able to achieve flow, but birders learn these skills as part of a "team" on guided bird walks. The lessons birders learn on the guided bird walks also form the basis for the naturalist gaze.

2 THE NATURALIST GAZE

ON A CRISP, early summer morning in late July, a group of ten birders followed the bird walk leader down a heavily wooded path. Walking in single file, we crept along as quietly as possible so we could hear any bird calls. Everyone's hands were on their binoculars, ready to focus on a bird at a moment's notice. We had not seen many birds that morning. In summer, birders usually hear more birds than they see. The trees are full of leaves, which easily obscure any birds perched on a tree limb. Summer means nesting time for birds, and their nests are often hidden, off the main path, to avoid any potential predators. In summer, birders take what they can get. That morning, we had heard a Catbird, a Blue Jay, and a Yellow Warbler. While walking through the parking lot to get to the path, we had seen Rock Pigeons, House Sparrows, and Red-winged Blackbirds feasting on some bread on the ground. We were frustrated. When we got to the end of the path, which opened out onto a large pond, we spotted two majestic Mute Swans. Everyone stopped to admire the large, white birds. "Ooh, look at them!" a woman exclaimed. Previously silent, the group now murmured in their excitement.

The walk leader quieted the group with some unsavory information about Mute Swans: "That's a nonnative species. They were brought to Long Island for the fancy estates, because people wanted their estates to look majestic. Mute Swans are very aggressive—they will kill other birds and other animals. They're not afraid of humans, either! They will come after you!" He then held up his hands, fingers spread wide, and said, "Their feet are this big." We now looked upon these birds with a new set of eyes. Instead of majestic, they now looked frightening and dangerous. We had learned that they *were* dangerous—to people and to other birds.

When I first started participating in bird walks, I didn't understand why birders admired some birds and not others. We spent many a morning fawning over relatively nondescript, small brown birds, or "little brown jobs" as birders liked to call them. As a novice, I couldn't see why the seemingly mundane brown birds received more attention than the charismatic, strikingly white swans. After this lesson on Mute Swans, and several others like it, I realized the walk leaders were encouraging us to develop a "naturalist gaze." Birders learn to view birds—as well as plants, animals, and humans—through a particular lens, which helps birders understand the interconnected relationship of all these beings in their appropriate ecosystem. The bird walks described in the previous chapter help birders develop their naturalist gaze, and this chapter explains the central elements of the naturalist gaze.

At its most basic, birding involves looking at birds. However, birders do not simply watch birds in any old fashion. Birders develop what I call the "naturalist gaze," which comes from watching wildlife through the lens of seeing what is in the best interest of native wild animals. The naturalist gaze instructs birders to appreciate healthy wild birds in the wild, in their natural habitat, with minimal human intervention. The naturalist gaze does not exploit or covet birds; birders appreciate catching even a momentary glimpse of wild birds in their natural habitat. The naturalist gaze brings together several elements of birding and other naturalist activities.

The naturalist gaze is *informed* by field guides, by scientific research, and by environmental and wildlife conservation information. To this extent the naturalist gaze is historically bound; as scientific knowledge grows, such as our understanding of the causes of extinction, the naturalist gaze evolves. The naturalist gaze is *evaluative*: it judges the flora and fauna in an ecosystem on the basis of this scientific information. The naturalist gaze is *concerned* about the health and well-being of wildlife in its natural habitat. The naturalist gaze is *integrative*: it views flora and fauna as interconnected elements in an ecosystem, which includes humans. The naturalist gaze therefore is not anthropocentric, and it is solely concerned with wildlife—not with companion animals or farmed animals. The naturalist gaze is *instructive*: birders use it as a way to teach others how to watch wildlife. Finally, of course, the naturalist gaze is *pleasurable*: people enjoy watching wildlife in its natural habitat.

The naturalist gaze differentiates itself from other types of gazes that people use to look at animals. The scientific gaze looks at animals in captivity, typically in laboratories, and sees animals as objects of analysis rather than as beings who might possibly prefer freedom.[1] Similarly, the disciplinary gaze can be directed at farmed animals, when farmers scrutinize and control animals' every movement on factory farms.[2] The zoological gaze describes how people look at captive animals in zoos, primarily for entertainment, but also for instructional reasons.[3] The zoological gaze represents a slightly more educational version of the tourist gaze, through which people also watch captive animals for entertainment.[4] Each of these gazes is anthropocentric, focused on the pleasures or interests of the humans who are watching, and not on the needs, desires, or interests of the animals being watched.

The art historian John Berger's classic essay "Why Look at Animals?" exemplifies this anthropocentrism. To Berger, looking at animals primarily helps us better understand ourselves. When we look at animals, we do not see them as beings with their own interests and desires. We see ourselves in them, we think about ourselves while we look at them, and their look comforts us: "With their parallel lives, animals offer man a companionship which is different from any offered by human exchange. Different because it is a companionship offered to the loneliness of man as a species."[5] Of course, people do not necessarily look at animals solely or even primarily to understand themselves; Berger's argument here simply retains some of the anthropocentric vestiges of academic understandings of nonhuman animals.

In contrast to such anthropocentric views, the naturalist gaze prompts birders to think about the birds and their thoughts, as Chloe described: "You imagine yourself thinking, 'What are they thinking?' They look at you with that one-eyed look thing, and you have to think that they see you, and what are they thinking?" Chloe doesn't describe seeing herself or better understanding her connection to nature when she looks at a bird and the bird returns her gaze. Instead, when Chloe looks at a bird, she turns her focus to the bird and wonders what the bird is thinking. Birders center the birds and their needs when they look at them and contemplate them, rather than using birds as a foil for better understanding "human nature" or themselves. This centering of the birds and their needs, and decentering humans and our needs, differentiates the naturalist gaze from other ways of looking at animals. As this chapter shows, the naturalist gaze highlights the centrality of natural habitats for birds, teaches birders about birding ethics through conversations with other birders and by learning about birds' needs, and provides a mitigating force against the excitement of chasing rare birds or misidentifying birds.

THE "AURA" OF BIRDS

People have always watched birds, and they have always tried to capture the images of the birds they have seen. In 1820, a young John James Audubon declared his intent to paint all of the birds of North America, and he published his illustrations of 435 birds in *The Birds of America* in a series between 1827 and 1838. Unlike the dead birds that populated still-life paintings of the time, Audubon painted birds in their natural habitats, such as particular species of trees. Audubon's lifelike illustrations belie the fact that he shot and killed birds by the dozens for each painting he completed. To give the appearance of realistic movement, Audubon held the dead birds in lifelike positions using wires. Early field guides such as Florence Merriam's 1889 *Birds through an Opera-Glass* contained wood engravings of birds as illustrative decorations, and John B. Grant's 1891 *Our Common Birds and How to Know Them* used black-and-white photographs of stuffed dead birds as an attempt to aid in identification.[6]

Now, bird photography is serious business, and many birders own photography equipment that vastly outpaces their binoculars in both magnification and price. Even with the developments in the clarity and accuracy of capturing bird images with modern photography, anyone who sees a painting or photograph of a bird they have seen in real life will remark on the difference between those images and the "real thing"—the living, breathing bird in the wild. The cultural theorist Walter Benjamin called this ineffable substance the "aura."[7] Anyone, birder or not, can intuitively understand this phenomenon when they think about the difference between seeing a photograph of the Grand Canyon and seeing it in real life. The naturalist gaze deepens our understanding of the aura by highlighting the importance of natural habitat to birds, and to watching birds.

Although Benjamin discussed the aura primarily in relation to art, his basis for the concept was nature. Benjamin described viewing a mountain range in the distance, or even sitting under the shadow of a tree branch, as a way of experiencing the aura of these majestic natural phenomena. To Benjamin, the mechanical reproduction of art that was designed for reproducibility, such as film or photography, killed the aura found in original, "authentic" works of art. Birders would agree—a photograph of a bird does not capture the same aura, or essence, of seeing that bird in person, in the wild. People attempt to reproduce natural objects in photographs or videos, and these reproductions differ from what we see with our own eyes. In attempting to reproduce the aura of natural phenomena, we lose its uniqueness and intransience.

In watching birds, birders appreciate what Benjamin called the "quality of its presence," or birds' aura as wild animals.[8] Some birders photograph birds for identification purposes or for enjoyment as an artistic hobby, but the primary goal of birding is to observe birds in their natural habitats. On walks, birders experience a wonder in seeing particularly beautiful or interesting birds. Murmurs of admiration and even literal "oohs" and "aahs" pepper each bird walk when we see a spectacular bird. Birders recognize this phenomenon and even cite "oohing and aahing" over a bird as a regular facet of a bird walk. In describing a typical day of birding, Vivian started by talking about how she plans the walk, how she starts her day, what she eats for breakfast, and the like. Then in describing the walk, she said, "Lots of walking, lots of talking, oohing and aahing over whatever we see, showing things. I also like just being outside, and we'll stop for everything: bug, butterfly, pretty flower, whatever."

As I said at the beginning of this chapter, when I first started participating in bird walks, I didn't understand why birders admired some birds and not others. I learned that large, majestic Mute Swans did not garner the same attention as small, seemingly nondescript brown sparrows, because of the naturalist gaze. Despite their size, beauty, and elegance, Mute Swans are not native to the United States, and their aggressive behavior puts native birds and local ecosystems at risk. Each time a participant—usually a new birder—expressed admiration for Mute Swans, the walk leader explained that Mute Swans are vicious, aggressive, invasive, and even dangerous to humans. Walk leaders train participants to view nature through the naturalist gaze, and thus they discourage participants from admiring charismatic and beautiful birds that are invasive, nonnative species. I explore birders' evaluation of nonnative species in more detail in chapter 5, but this example demonstrates how the bird walk leaders did not see the "aura" of Mute Swans. As nonnative species, they did not hold an authentic aura of wild animals in their natural habitat.

This distinction between birds deserving of our admiration and those that do not demonstrates a key way that birders deepen Benjamin's concept of the aura. Benjamin clearly did not describe the natural aura as a naturalist. To birders and others viewing nature through the naturalist gaze, not just any tree limb would demonstrate the natural aura. If the tree limb Benjamin described came from an invasive, nonnative tree species, it would not garner wonder, attention, or admiration—it wouldn't have the natural aura. The former philosophy professor turned wilderness guide Jack Turner wrote, "The loss of aura and presence is the main reason we are losing so much of the natural world."[9] Birders sense this loss as well, which is why birders explicitly align the aura of nature to the naturalist gaze when they admire native, noninvasive bird species in their natural habitat.

The naturalist gaze extends to other elements of birds' habitat, such as plants. Birders disdain nonnative plants because they harm birds' habitats. Birds eat the seeds and the insects found on native plants, and nonnative plants do not provide such sustenance. Worse, many nonnative plants are invasive, meaning they take over the habitat, killing native plants and eliminating birds' food sources. I took a walk with a local Audubon chapter on a bracingly cold January morning that was entirely focused on tracking the invasive plants in their sanctuary. We enjoyed seeing some birds along the way, and everyone had their binoculars strapped to their chests, but the majority of the walk entailed taking side trips off the main path to closely examine particular plants and to verify their species. The two main leaders kept griping about how the neighboring golf course as well as the local electric company, whose power lines cut through the sanctuary, weren't helping battle the invasive plants. Despite this walk's focus on eradicating invasive plant species, one of the participants kept remarking on the beauty of invasive plants. "But they're so beautiful!" she'd say each time someone complained about an invasive plant's growth. Each time she did this, the group yelled at her—at first jokingly, then a bit exasperatedly, especially when she picked some invasive phragmite to take home with her. Thus, even if a plant species were beautiful, it would not have an aura and would not garner admiration if it were an invasive, nonnative plant species.

In his discussion of the aura of original works of art, Benjamin referred to a work's "presence in time and space," explaining that "the presence of the original is the prerequisite to the concept of authenticity."[10] Benjamin went on to argue that natural objects also have an "aura," but he mentioned only mountains and trees in his example. How might living creatures such as birds express this element of the aura, this presence of originality? We don't have "original" or "authentic" birds, whose aura was destroyed by mechanical reproduction. Instead, Benjamin argues that natural objects' aura comes from the "unique phenomenon of distance, however close it may be."[11] In one sense, he literally means distance, like "a mountain range on the horizon."[12] We can also experience this aura from natural objects that are physically closer to us, like a butterfly that alights on our hand, a fox that crosses our path, or a bird we are able to observe at close range (or even closer, with the use of binoculars). Thus in another sense, he means distance as in otherness, their distance from our lives as people living in built environments. Part of birds' otherness and aura comes from their mobility, which distinguishes birds from immobile mountains or even from flexible trees. Birds' ability to fly contributes to their "presence," or their charisma, as beautiful, interesting living beings.

Cynthia works as an education manager for a local Audubon chapter, but she originally went to school for entomology. She explained to me that since birds are more charismatic than insects, they're much more useful for getting people to care about conservation issues, which include insects: "If you want to accomplish entire habitat restoration or support ecosystems, birds are a really important way of doing that. I'm really glad it's become a bigger part of the conversation, how birds rely so much on insects, and it's getting people to be much more aware of the importance of our creepy crawlies." Birds benefit from people's positive associations, as opposed to the negative symbolism of insects and other "creepy crawlies." Birds' charisma promotes larger conservation issues to people already interested in birding, as Cynthia explained that birds are "definitely a great tool" for that.

Birders frequently employ the National Audubon Society's motto of "connecting people with nature" as a way of getting people to protect the environment. When I interviewed Chloe, who presents Audubon programs and materials throughout rural Wyoming, she explained that anyone who is "passionate about birds" will connect birds to their natural environment, including "the habitat that that bird needs, the food that bird needs, its part in the food web, what it eats, what eats it, and when you learn all of those components about those things, I think then you're just drawn to want to protect it." Chloe's educational background is in aquatic biology, but she also found birds to be more useful than fish for promoting conservation issues: "Birds are so easy to identify because they do happen to be charismatic, and they're bigger, and people like to see them, and because of that, I think people are drawn to protect areas where they are, especially if they go there all the time. They wouldn't care about some random pond being filled in for a housing development to be put on it, unless you go to that pond all the time to see those birds. You see those birds there, and then you say, 'Wait a minute, you can't do that, because that's where those birds are.'" Birds' charisma creates part of birds' aura. People become interested in birds because they are charismatic, which then leads people to want to protect birds and their habitats. As Chloe said, people won't care about a "random pond" getting filled in for a housing development, but they will care about a pond where they go to watch birds. Thus, birds' aura relates to caring about birds' habitat, and the naturalist gaze encourages birders to maintain that habitat.

Birds' otherness and aura in their ability to fly also create a challenge to birders attempting to see a bird, which can fly away at any moment. Birders relished this challenge, as it helped them enter into a flow state, as described in the previous chapter. Flow activities must be challenging enough to maintain attention, but not so challenging that the activity becomes frustrating.[13] Miranda compared *birding* with bird *watching*, and she emphasized that birding represents the more active, challenging activity: "Well, they move around [*laughs*]. That's sometimes a challenge. I think it's pretty active. People talk about bird *watching*, and to me that feels static, like you're staying in one spot, just watching the birds as they come to you. But really, *birding* is more active, it's outside, and you're walking around, and you're seeking out birds, or seeking out new habitats to see what birds are in them, that you might not get so much if you were just really into moss. Maybe, maybe not. I don't know—I'm not really into moss." To Miranda, birding presents more of a challenge than bird watching because in bird watching, the observer sits in one place, usually a backyard, and watches the birds as they come to her. In contrast, in birding, the observer follows the birds, walking through different habitats and actively seeking out birds, which can fly out of range at any moment. Birders thus argued that birding is more challenging than other naturalist activities, as Mona explained when contrasting birding with identifying wildflowers: "The challenge. It's more challenging than a flower. It's not going to go anywhere. A tree's not going to go anywhere. But you have that brief moment, that flash of seeing that bird, that, if you really want to have a good day with a lot of birds, and you really want to ID them, it can be gone in a flash, and you may never see it again. It's challenging. It's more challenging than other naturalist activities."

To Benjamin, natural elements such as trees or mountains possessed an aura because of their distance and their otherness. To birders who watch mountains or trees through the naturalist gaze, not every mountain range or tree, or even other aspects of nature, would hold an aura. Invasive, nonnative tree or bird species would not have an aura, and even native trees might not have the same aura, since birders saw less of a challenge to identifying them. Birds' ability to fly away, out of sight, differentiates birds from mountain ranges, Benjamin's example of choice. Part of the wonder, the aura, of seeing birds was that birds could simply fly away, never to be seen again. Thus the challenge of birding, and birds' ability to fly, added to the aura of birds. Simply getting the chance to spot a bird before he or she flies away was one of the joys of birding.

Chloe emphasized that she enjoyed the challenge, but she also talked about how exciting it was to see anything, whether a rare or common bird: "It's kind of like a *Where's Waldo*, you know, but in real, 3-D, real time. You might hear them, you don't always see them, but you know that they're there, and you go through the process of, you never know what you're going to find. It's always an adventure. No matter where you are. It's exciting when you see something. I'm not particularly snobby as far as it has to be some rare bird, or it has to be something I don't see all the time, but that does always up the ante just a bit. Just seeing life is fun." Chloe's description exemplifies the pleasurable and informed aspects of the naturalist gaze. Birding is a fun challenge, an "adventure" where you never know what you will find. This aspect further differentiates birding from watching wild animals in captivity, where no element of surprise exists and the animals are removed from their natural habitats. Birders' appreciation of watching birds in their natural elements, free to move about, makes up a significant portion of the naturalist gaze.

DEVELOPING WAYS OF SEEING THROUGH TALK

One of the ways that birders learn which birds they should expect to find is by talking to other birders about birding in specific areas. In his book on mushroom collectors, the sociologist Gary Alan Fine explained that since mushroomers do not necessarily share any background other than a love of mushrooming, much of the talk when they gathered for their meetings, beyond the meeting business itself, centered on stories of mushrooming.[14] Birders, too, talk about birds they have seen as a way of generating or solidifying group culture. Fine's mushroom collectors' talk primarily came from newsletters, meetings, or interviews. In contrast, the birders I observed primarily talked about birds they had seen while in the field, watching other birds. Ostensibly, they were in the field to look at birds, but when birds were not around, birders wistfully told stories about exciting birds they had previously seen. Birders also told stories while at meetings, as did Fine's mushroomers, where each meeting would begin with birders sharing their "sightings" from around the area. What birders do, that mushroomers do not, is spend their downtime while looking at birds talking about, well, other times they have looked at other birds. While out in the field birding, birders talked about birds they had seen at the same location on previous trips, at other local places, or even earlier in the day, at the same location.

Talking about birds previously seen at the same location helps develop birders' naturalist gaze. If birders know which birds they should expect to see in a given location, they learn more about the birds' habitats, and they hone their ability to search for and identify the birds in the area. For example, one of the groups I observed often walked in a park that had two lakes. Every time we got to the first lake, the leader, or other group members if it was a new leader, would talk about Green Herons. The leader would tell us to scan the edges of the lake and look for Green Herons, or the group members would tell new leaders that we had previously seen Green Herons at this lake. Thus, group members were learning to look out for Green Herons whenever they approached this particular lake.

In a second form of talk, birders talked about birds they had seen at other local places. In the New York metropolitan area, these most often revolved around where to find eagles. Bald Eagles had made a successful comeback in the area, often nesting in the area as well. Teatown, a nature center in Westchester County, hosts EagleFest every year to celebrate the eagles' return to the lower Hudson valley. On birding walks, people often talk about their favorite spots to find eagles' nests or to see eagles feeding in the Hudson River. On one walk, two people discussed different spots where they had seen eagles' nests—one near a prominent local building and one near a particular lake—and another person told a story of a nest that had fallen down and the eaglets had to be rehabilitated. Again, these stories hone the naturalist gaze, so birders will know which birds to expect when they visit those other local areas.

The third type of talk consisted of birders sharing birds they had seen earlier in the day, at the same location. This most often happened when birders crossed paths with another group of birders. Each time they passed other birders, the groups asked what the others had seen, and the birders readily shared species and location. Sometimes birders shared photographs of interesting species they had taken earlier in the day. Unlike Fine's mushroomers, who developed a code of secrecy, these birders readily shared information on where to find specific birds.

Why do birders share information that mushroomers try to hide? First, Fine's mushroomers believed people had to put forth a certain amount of effort in order to fully enjoy their finds. I found no such justification for withholding information among birders. Birders readily share where and how to find birds. Their only encouragement for new birders is to study the birds' diagnostic markings so they will improve at identifying the birds when they see them. Second, Fine's mushroomers worried about other mushroomers overpicking rare mushrooms, thus ruining the opportunity for others to enjoy them. There exists no equivalent concern for birders, since the same bird can be seen by multiple people. Birders discourage sharing nesting locations because disturbing nests can harm birds, and this protection of vulnerable birds makes up part of the naturalist gaze. Walk leaders reminded participants of this, as Luis did on one walk when we started discussing owls.

"If you ever find an owl's nest, you don't report it publicly. You only want to share it with other ethical birders," Luis said.

"Are there any owls in this park?" someone asked.

"Yes," Luis responded.

"I have a friend in Queens who found an owl's nest, but he wouldn't tell me where it was, because he said people would try to steal from the nest."

"Or get too close to the birds," Luis added.

"I asked him for a couple of weeks, and he finally took me out there."

"Where?" Luis asked, and the man responded with the name of a park in eastern Queens.

"So your friend wouldn't tell you, but now you're telling all of us?" Luis half jokingly, half admonishingly responded, and they both laughed.

Even birders participating in big day contests, like the World Series of Birding, readily share location information. At the finish line of the World Series of Birding, teams all shared exactly where they had seen each bird. One birder showed me a smartphone app developed for the contest, where teams could share locations and weather information. In this friendly competition, the birders shared information so that everyone could see as many birds as possible, stay dry, and have fun.

In my interviews, birders explained that this sharing culture represented one of the aspects they enjoyed about birding, although not everyone readily participated in it. When I interviewed Mona, an accountant, at the National Audubon Convention, she told me she liked the "socialness" of birding, and she explained:

> You seldom meet somebody who's not open and friendly and willing to share. There are some egotistical, usually men, out there, a few women, sometimes, that we run into that are a little egotistical and not really open and friendly, but the sharing, the sharing of the stories. You can always ask people, if they say, "I was in Vermont last month," you can ask, "What kind of birding was there? What exactly did you see? Where exactly did you go?" Or if they say they saw a specific bird, you could turn around and ask them, "Where did you see it?" It might be a life bird for you, and it might be close by, and they're always willing to share.[15]

I frequently witnessed this culture of sharing on bird walks. Whenever the group encounters other birders, people ask, "What have you seen?" or "Seen any good birds?" and they share information about exactly where they had seen particularly interesting birds. In this sense, birding diverges from other secretive subcultures. Deviant or illegal subcultures like ravers or drug smugglers maintain strict secrecy about their activities, for fear of being arrested.[16] Other subcultures associated with nature maintain such secrecy as well, such as mushroomers, who want to avoid overpicking mushrooms, or surfers, who want to keep their "secret locations" for good surfing to themselves, so the beaches don't become overcrowded.[17] Because birders do not have structural reasons encouraging secrecy, such as fear of arrest, overpicking, or overcrowding, they enjoy the cultural aspects of sharing their "finds" with an appreciative audience of other birders.

THE NATURALIST GAZE AND BIRDING ETHICS

With its focus on bird habitats and biological needs, the naturalist gaze also informs birding ethics, which instruct birders in how to engage with birds in the wild. In her discussion of the destruction of a Cattle Egret nesting colony in Conway, Arkansas, the environmental sociologist Stella Čapek argued that people should consider "bird space" and "bird time."[18] Developers building a new residential neighborhood plowed into the Cattle Egret nesting colony during the nesting period of June, without thinking about the fact that June means nesting season for birds. After birds' spring migration, they settle down to build nests and raise babies, often returning to the same nesting spots year after year. Had the developers respected "bird time" and understood that June was a critical time period for birds, they might have delayed their work until after nesting season. Or, had the developers understood "bird space" and realized that this plot of land served as nesting ground for Cattle Egrets year after year, they could have moved their development elsewhere. Čapek focused on nesting grounds to exemplify bird space, and breeding season for bird time. I take these concepts further to show how birders hold an even deeper understanding of bird space and bird time, and how such understandings inform the naturalist gaze.

Birders live their lives by "bird time." Each season brings a different activity to observe, and the most exciting times of the year for birds become the most exciting times of the year for birders. In winter, birders typically observe "permanent residents," or birds that don't migrate because they have plenty of food sources year-round. Other birds engage in spring and fall migration so as to find food and nesting locations. These migrations can range from short distances between a few states to quite long distances between continents. Long-distance migratory birds breed in the United States or Canada in spring and summer, and in fall they migrate down to spend winter in Central or South America. Spring migration and breeding periods are the highlight of the birding year, since birders can see rare migratory birds passing through on the way to their nesting grounds, and they can observe the courtship patterns and bright breeding plumage of the birds that nest in their area. It's no happenstance that most birding festivals occur in spring. Birders know exactly which time of the year they should expect to see particular birds, and since many birders keep notes of the birds they see in their backyard, they know whether they see birds earlier or later in the year. Birders' sense of "bird time" thus informs their naturalist gaze.

Birders also understand "bird space." One very important space for birds is their nests. Bird time tells birders that spring and summer are nesting season, and respecting bird space means building nest boxes for birds to use, not disturbing nesting birds, and knowing whether and how to help a baby bird found out of his nest. Nonbirders often think that baby birds out of their nest are in danger when they are not. Birders and birding organizations share information on fledgling birds throughout late spring and summer, so as to avoid unnecessary human intervention (which may end up harming the birds more than helping them). Another aspect of bird space is their habitats, which include nesting areas, in addition to providing birds' water, shelter, and food sources. We typically define habitat by the types of plants in the area—cypress swamp, grassland, pine forest—but habitat also includes the landforms, climate, and other animals living in the area. Birders use habitat to provide a clue as to which types of birds they might see in the area. Birders also work to improve birds' habitats by planting native plants in their own yards and encouraging cities to plant native plants in parks and other public spaces.

To "bird space" and "bird time" I would add "bird senses," including vision, behavior, and hearing. Birders understand what a bird's-eye view of the world looks like, and this knowledge informs how birders help birds maneuver their flight paths. Bird strikes—collisions with glass—represent birds' single largest killer. Hundreds of millions of birds die each year when they collide with glass on buildings, especially in cities. Birds also become trapped and disoriented in artificial light at night. Scientists don't yet understand exactly why birds get trapped in light, but they do know that birds can't differentiate the solidness of glass from an unobstructed flight path. Birds can crash into glass if they see plants or trees behind it or reflected in it. Birders must understand bird vision in order to help birds safely navigate humans' built environments. Birds have much larger eyes than mammals—they only appear small because all but the pupil is covered in skin and feathers—which means birds have much better vision than mammals. Further, whereas humans have three photoreceptors, or cones, in the retina, birds have four—red, blue, and green, like humans, but also ultraviolet cones. Birds use their left and right eyes for different tasks, and they sleep with one eye open. Most importantly for bird safety, birders understand that birds either cannot detect windows or misinterpret them. Birders have created window decals to put on windows so that birds can discern and avoid them. This problem becomes worse at night, when most birds make their migrations. Birds use the light from the setting sun, moon, and stars to navigate their flight paths, and thus they become disoriented by artificial light on buildings. Many Audubon Society chapters campaign for cities to turn off building lights at night during migration seasons, and New York City Audubon's Project Safe Flight monitors the 9/11 Memorial "Tribute in Light" each September so that organizers can turn off the lights periodically to let trapped birds continue their migration. Thus birders' understanding of bird time, bird space, and bird vision is informed by science, which informs the naturalist gaze, which informs birding ethics.

All gazes hold an element of power. Those doing the gazing typically hold some form of symbolic or material power over those being gazed upon. What differentiates the naturalist gaze from other types of gazes is that the naturalist gaze helps birders understand the power relations in watching wildlife. Birders' understanding of bird time, bird space, and bird vision allows birders to engage with birds in what they see as an ethical approach, in ways that provide the most benefits to native wild birds and that minimize human impacts on those birds' well-being. The naturalist gaze thus informs birders' ethical guidelines for watching birds in the wild. These ethics describe practices that birders engage in on their own, as well as practices that birding organizations explicitly describe and prescribe, such as the American Birding Association's code of ethics for birding:

1. Promote the welfare of birds and their environment.

1(a) Support the protection of important bird habitat.

1(b) To avoid stressing birds or exposing them to danger, exercise restraint and caution during observation, photography, sound recording, or filming.

Limit the use of recordings and other methods of attracting birds, and never use such methods in heavily birded areas or for attracting any species that is Threatened, Endangered, of Special Concern, or is rare in your local area.

Keep well back from nests and nesting colonies, roosts, display areas, and important feeding sites. In such sensitive areas, if there is a need for extended observation, photography, filming, or recording, try to use a blind or hide, and take advantage of natural cover.

Use artificial light sparingly for filming or photography, especially for close-ups.

> 1(c) Before advertising the presence of a rare bird, evaluate the potential for disturbance to the bird, its surroundings, and other people in the area, and proceed only if access can be controlled, disturbance minimized, and permission has been obtained from private landowners. The sites of rare nesting birds should be divulged only to the proper conservation authorities.
>
> 1(d) Stay on roads, trails, and paths where they exist; otherwise, keep habitat disturbance to a minimum.[19]

Birders do not carry a laminated code of ethics with them for reference on their walks. Rather, participants in bird walks learn about birding ethics from walk leaders, in the moment, and immediately put them into practice. On one walk through a grassy field late in June, Arnie, the walk leader, pointed out a Killdeer on the ground. As we moved closer, we watched the bird awkwardly hold out his wings and start chirping. We all watched through our binoculars as this bird walked around, calling and dragging his wing on the ground. Once we saw this display, Arnie urgently told us, "This bird is pretending to have a broken wing, so he can lead us away from his nest." Arnie explained that Killdeer are shorebirds that adapted to grassy land, so we must be near the bird's nest somewhere on the ground. Killdeer conduct a "broken wing" act or distraction display to lure any potential predators away from their nests or hatchlings, and we were witnessing that act. He said we should very carefully walk back to the main road and use the footpath so that we don't bother the bird. "I don't want to keep going when the bird is clearly in distress," he said. "That would go against the birding code."

Even though this Killdeer was an uncommon bird, exhibiting quite interesting behavior, Arnie did not keep us there to observe the bird and his behavior. Instead, he put what he believed to be the bird's welfare first and led us away from the Killdeer and his nest so that the bird would be, Arnie hoped, safe and out of distress. Here, birding ethics were informed by bird time (late June is nesting season), bird space (Killdeer nest on the ground), and bird behavior (the Killdeer's distraction display). An uninformed observer may have simply thought such a display was interesting, but our walk leader was able to put all of these elements into context, using the naturalist gaze, and encourage what he believed to be the ethical behavior of leaving the bird alone. The American Birding Association code of ethics (1b) tells birders not to stress birds and to keep away from nests. The leader, of course, did not cite

FIGURE 2. Killdeer's broken wing display (photograph courtesy of Dave Saunders).

this particular code, but he effectively put it into practice in leading us away from the Killdeer's nest.

Birders also understand and use birds' sense of hearing to inform their ethical engagement with birds, as Diane explained to a group of birders participating in a workshop on birding by ear that she led. Diane, an experienced birder and naturalist, first explained to participants exactly why birds sing: to impress other birds (especially mates), to demonstrate their health, and to create a "sound fence" to defend their territory. Since birds have a double syrinx, as opposed to our single larynx, they can create beautiful sounds that humans cannot. Diane taught us about song quality, pitch profile, and song structure as various ways of differentiating among bird sounds.

At the end of the prepared lecture, we had a chance to ask questions. One of the participants asked, "Can't you play songs to attract birds?" Diane quickly responded, "Yes, but don't do it." She then went on to explain: "I'm tempted to do this when leading a walk, because I want everyone to be able to see the bird. But if you use playback during breeding season, or during winter, it will have effects on the birds, such as taking them away from their food source." What's more, Diane noted, "playback is prohibited in state and national parks, and it's prohibited during contests like the World Series of Birding. We have a link on our website to the American Birding Association's list of birding ethics, if you want to learn more."

With the proliferation of birding apps on smartphones, playback represents a growing ethical concern for birders. "Playback" refers to playing audio recordings of bird calls or bird songs to lure a bird out of hiding. This may at first just seem like an interesting trick for getting birds to move into the open for better observation, but birders, using their naturalist gaze, recognize the ethical quandaries involved in such playback. Consider, for example, the Kauai 'o'o, a Hawaiian honeyeater that scientists declared extinct in 1987. The last bird of the species was a male, and ornithologists recorded him singing his haunting, flute-like mating song for a female bird who would never arrive. Imagine if some overzealous person, intent on seeing this last bird, played the call of a female bird to lure him out of the forest. Luckily, as far as we know, that never happened. With more species becoming threatened, endangered, or of special concern, the American Birding Association's code of ethics stressing to never use such methods for such birds makes all the more sense.

Finally, bird photography shows how the naturalist gaze informs birding ethics. The American Birding Association's code of ethics tells would-be photographers to "exercise restraint and caution" so as to avoid putting birds in danger or under stress. While photography is not the focus of the birding walks, many participants bring their cameras, and sometimes tripods, to photograph the birds observed on the walks. The birders I interviewed spoke about the need to follow such rules, rather than just trying to get the best shot. When I interviewed Tom, a retired carpenter who goes on nature walks every single day, rain or shine, he complained about "overzealous birders" who trespassed on private property to get a photograph of a rare Gyrfalcon. He worried that this would "give birders a bad name," and that each time people see a birder they will think poorly of birders. But more than that, he worried about how unscrupulous photographers affect the birds: "We do the January trip, the first day of January, to Jones Beach, and a couple of years ago, there was a Snowy Owl there. And the park had it roped off, just to leave the thing be, and these people, they got telescopic lenses, anyway, but they have to get closer, they have to climb over the barricades and go, and then spook the bird." Tom was upset because overzealous bird photographers frightened and agitated the birds in their attempts to get a good photograph. The Audubon Society also encourages birders to engage in ethical bird photography by not luring owls with mice to get a shot, and not using a flash when photographing owls, since the bright light can harm their sensitive eyes.[20] All of the birders I interviewed, and the leaders on every bird walk I attended, emphasized birding ethics, even at the expense of seeing a bird. The welfare of the bird always came first. Therefore, the naturalist gaze means that sometimes birders do not get to gaze upon a bird at all, because the goal is to not harm the bird.

EXCITEMENT CLOUDING THE NATURALIST GAZE

The naturalist gaze represents the "ideal" way of looking at birds. Birders want to always keep birds' best interests in mind, and they often stop themselves from watching rare birds in the field if they think they might endanger the bird. Sometimes, the excitement of seeing a rare bird, in the moment, in the field, clouds the naturalist gaze. Even though birders know that birds out of their range are in danger, they still get excited about the possibility of seeing them. Birders also know that accurate species identification is crucial for citizen science and other conservation projects; however, the excitement of a possible rare species can cloud their judgment and encourage misidentification.

This difference between the "ideal" and "excited" ways of looking at birds resembles the social psychologist Steven L. Gordon's discussion of the differences between institutional and impulsive emotions.[21] We understand the meanings of emotions by linking them to institutionalized meanings or to more spontaneous, impulsive expressions. The naturalist gaze represents an institutionalized emotional response to birds. It is institutionalized formally in codes of ethics, and informally in conversations on bird walks. Novice birders exemplify impulsive emotions when they excitedly react to the majestic beauty of a Mute Swan. Bird walk leaders, embodying institutionalized emotion in the ideal form of the naturalist gaze, will quickly inform the group that Mute Swans are vicious, dangerous, and harmful to other birds, as well as nonnative species, and thus they temper those initial excited responses. Even experienced birders fall prey to impulsive emotional reactions when they see rare birds. This excitement and impulsive reaction happen only in the field when rare birds are right in front of them. Otherwise, birders can stave off these impulsive emotions and maintain the institutionalized emotion of the naturalist gaze.

Birders can sign up for "rare bird alerts," in which they receive emails, texts, and even phone calls alerting them to a rare bird in a specific county, state, or region. Many birders travel to see rare birds; this is a regular part of birding, especially for birders on the more competitive end of the spectrum, such as those engaging in "big years." As I observed and interviewed less competitive and more conservation-minded birders, they said the opposite: they specifically do *not* travel to see rare birds. Since the naturalist gaze encourages watching birds in their natural habitat, these birders did not wish to see birds that are lost, out of place, or in a habitat where they do not belong. Seeing such birds through the naturalist gaze means knowing that birds out of their habitat are more likely to be in danger and more likely to lack appropriate food or shelter. When I interviewed Lenny, a retired biostatistician, he explained that he did not like to "twitch," or chase vagrant birds that are out of their natural habitat: "In their native habitat. That's why a lot of times we don't chase exotic or unusual birds. There's a Ross's Gull up now in the Adirondacks. It doesn't belong there. I'd rather see them up in the Arctic where they belong, rather than being kind of lost and not knowing how to act." Birders can avoid traveling to see rare birds, because they have to take the time to plan and execute the trip. That gives the institutionalized emotional response of the naturalist gaze plenty of time to remind birders that it's probably not a good idea to travel just to try to see a bird that is completely out of his element. The case of a rare Painted Bunting that showed up in Brooklyn demonstrates this response.

In late November 2015, a Painted Bunting appeared in Prospect Park in Brooklyn. The range of this beautiful multicolored bird typically encompasses Central America and the most southern parts of the United States. This rare and wildly out-of-range appearance captured the attention of birders throughout the New York area. Thousands of birders gathered in Prospect Park to see the bird, who remained there until early January 2016. Before a walk in early December 2015, a group of birders waiting to begin a walk discussed the bird.

"Have any of you gone to see the Painted Bunting?" a man asked.

"I wouldn't want to go out there with thousands of people," Brandon responded.

"I'm worried the bird's going to get trampled!" a woman said.

Later, when everyone had arrived and we started the walk, the discussion resumed. Without even specifically referring to the Painted Bunting, Tom, who arrived later, began the discussion:

"Did any of you go down to Brooklyn?" Tom asked.

"No," the first man to ask about the bird responded.

"Is it just a rumor?" a woman asked.

"No, there's video of the bird," Tom replied.

"There's probably 3,000 people out there, and I don't want to go out with all those people," Brandon added.

"I'd go, if I didn't already have this bird, but I've seen one before in North Carolina, which is about the northernmost that you'd find this bird," Tom said.

Tom's response demonstrated that he understands the pull of wanting to see such a rare bird, but since he had already seen a Painted Bunting in his natural habitat, he was not tempted to travel to see the rare bird. That Tom could ask, "Did any of you go down to Brooklyn?" without referring to the Painted Bunting also demonstrates that birders are so up to date on these issues that he could simply ask about Brooklyn and everyone knew he was referring to the Painted Bunting and not to the latest exhibit at the Brooklyn Museum.

The naturalist gaze's focus on seeing birds in their natural habitat does not prevent birders from traveling to see birds. Many of the birders I interviewed take birding tours or vacations specifically planned for the purpose of birding. Birders with the naturalist gaze focus on visiting birds in their natural habitats, rather than traveling to another part of the state to see a vagrant bird out of his normal range. They often talk about choosing tours that benefit the local region or the local economy and how that helps balance the environmental costs of their travel. Traveling to see just one rare bird is more difficult to justify, as Trish, a retired microbiology professor, said:

> I think there's some conflicts, you know, conflicting feelings regarding some birding practices. I do like to travel, and I do, for example, going to the Galapagos, it's definitely to see the birds. Well, that's not without a cost to the carbon situation. Although I think it helps protect the Galapagos for people to, for Ecuadorian people to see that that's an important resource. But things like, do I chase that bird? There's this rare bird alert, you could drive to such and such a place and see that bird, and I'm thinking "Nah." Is that really the best idea? Is it worth it? To add that bird. To say you saw that bird, in that year, in that county. For me it's not, but I know for some people that is an important thing.

Trish went on to explain that she was more intrigued by the possibility of doing a big year in her own county or own backyard than she was in traveling the world to complete an international big year.

In the time it takes to plan a trip to another part of the state to see a vagrant bird, birders can stave off the impulsive response to travel for just one bird. They can maintain the element of the naturalist gaze that focuses on seeing birds in their natural habitat. It is harder to resist an emotional response in the field, in the moment. Since most of the birders I observed and interviewed submitted their species lists to eBird, which uses those lists for citizen science projects, the birders knew that accuracy in species identification was crucial to that endeavor. However, the frustration of seeing a small number of birds, or the excitement of the possibility of seeing a rare bird, can lead to misidentification or wishful thinking, which can cloud the naturalist gaze.

For example, on a walk on a bitterly cold February morning, I joined a group of birders to look for birds by a frozen lake. When we arrived at the pier, we encountered another group of birders who were already there, and we all took turns spotting birds. One of the men in the other group said he thought he saw a pair of Mute Swans at the end of the lake, and another called out that he thought the white tufts at the end of the marsh might be a Snowy Owl. The other group left, and my group stayed for a while longer. Later, as we left the lake, I pointed out the supposed pair of Mute Swans and asked Lenny and Catherine what they thought it was. They said, "That's not a bird; that's white bags of trash!" Instead of a Mute Swan, they said, "that's a 'wish bird'—you wish it was a bird!"

Birders often mistook bags of trash for birds, or quickly turned their binoculars to follow what turned out to be a falling leaf, which they sometimes called "leaf birds." Even tree bark can be mistaken as a bird, as when Warren said he thought he saw a Great Horned Owl in a tree, but it was only owl-shaped bark. Everyone laughed, and one person joked with him, "You thought you had the bird of the day, huh? Keep hoping!"

Birders recognize this threat to their naturalist gaze with the joking monikers of "leaf birds" and "wish birds," and they also openly discuss the need to be vigilant against such clouding or obfuscation of one's naturalist gaze. On one walk, Tara asked how to identify a certain bird, and then she said, "I'm an occasional birder now. I get things wrong and I have no pride about that." Tara said it's harder for her to identify things now, and she tries not to identify a bird as a specific bird just because she *wants* it to be that bird. She said now she asks herself, "What do I want it to be? And now, why is it not that?" Tom commiserated and said that he tries to turn

Mockingbirds (a common bird) into Northern Shrikes (a rare bird) all the time.

Despite her explanation of her strategy for minimizing wishful thinking and keeping her naturalist gaze clear, Tara fell prey to the lure of a potential rare bird sighting later in the very same walk, when she tried to turn a group of Brown-headed Cowbirds (common birds) into a flock of Cedar Waxwings (less common birds). She kept saying that she thought at least one of them was a Waxwing, because she saw his crown. People in the group kept asking, "Are you sure? I'm only seeing Cowbirds, maybe a Red-Winged Blackbird." Even I was skeptical, but because of my novice status as a birder, I did not say anything. Tara kept insisting that at least one of the birds was a Waxwing, and she kept referring to a crown that nobody else was seeing. After a few minutes of debate, Tara said, "There *are* a lot of smaller birds in with those birds," and then finally, "Oh well, I guess that was wishful thinking."

This wishful thinking in identification is so common that organizations within the birding world have developed several checks and balances to account for them. eBird will reach out to verify any unusual sightings that people report. Some of these unusual sightings can be attributed to novice birders simply misidentifying birds, but as I just showed with the clouding of the naturalist gaze, they can even be misidentified by more experienced birders who wished a rarer bird onto their list. Correctly identifying birds with a calm, reasoned, naturalist gaze approach is important when it comes to engaging in citizen science, which I discuss in chapter 6. Birders want to submit correct species lists to help conservation science, and misidentifying birds because of "wishful thinking" would mar that goal.

Even friendly competitions, such as the World Series of Birding, have rules that help institutionalize the naturalist gaze and stave off wishful thinking. The World Series of Birding has rules governing whether a bird can be added to a group's species list: "All birds tallied must be identified by at least two members of the team. Identifications made by just a single team member may not be counted."[22] Whether part of the rules or not, birders in the competition remember to check their emotions so as not to misidentify birds. At the end of the World Series of Birding one year, a group of eight serious birders, all of whom have written books on birding or lead birding tours, were discussing their finds.

"The highlight of the day was when he couldn't identify a bird!" a woman said, pointing to another man in the group.

"What?" everyone shouted, in unison. The man looked sheepish.

"Tell us more!" they said.

"I had a hard time getting on it, so I didn't want to count it," he explained. They kept pressing him.

"It was a shorebird," he admitted. They asked for more.

"It nests in trees," he added, giving them hints so they could figure it out. One man guessed the bird, presumably correctly, because the birder just smiled and said, "Well, maybe."

"He didn't want to identify it because it's so rare, and so special, that he wanted to be absolutely certain in his identification," the woman clarified. The bird never made it on to their list, since he was the only one who saw the bird, and the rules require that two people make the identification. More importantly, this birder wished to be absolutely certain about his identification even before sharing it with other birders. Part of this hesitation can come from pride, but I argue that it comes from the naturalist gaze.

Sometimes, the cloudiness of the naturalist gaze occurred because of simply misspeaking or mishearing a bird's name. At the beginning of one walk, a birder said he saw a Bohemian Waxwing on the way in to the park. Bohemian Waxwings live in northern Canada and Alaska and would be rare in the New York metropolitan area. Upon hearing this, the walk leader's eyes widened in surprise, and the birder quickly clarified, "No, not Bohemian, *Cedar* Waxwing!" The leader said, "Whew, when you said Bohemian, I was about to have a stroke!"

On another walk, a couple of men ahead of the group stopped and intently watched an area on the edge of the lake. Someone asked what they were seeing, and they replied, "Goldfinch." A woman excitedly asked, "BULL Finch?" which is a bird found in the West Indies. The man laughed and said, "No, GOLD Finch. If it were a Bull Finch, I'd be freaking out!" The woman jokingly responded, "Be careful, that's how rumors get started. The next thing you know, we'll see this sighting on the main page of the local Audubon website." Birders know that misidentifying a rare bird can lead to other birders wishfully seeing the same bird, so they carefully keep their naturalist gaze focused on birds more likely to be seen in the area.

Birders who continue to search for Ivory-billed Woodpeckers, a bird presumed to be extinct since its last confirmed sighting in 1944, may represent the ultimate example of wishful thinking. Since 1944, several reported sightings have come from credible birders and ornithologists, and people have not stopped looking for the bird. The larger birding and scientific community awaits verifiable proof of the birds' existence before officially removing their extinct listing. Geoffrey Hill, ornithologist at Auburn University, described the immediate (and masculinist) reactions among a group of birders who purportedly spotted an Ivory-billed Woodpecker:

> Nobody collapsed. There were no tears of joy. No hysteria. Tyler was clearly excited. I was relieved. The best birder on our team had gotten a clear look at the bird. Our ivorybill discovery had just moved from possible to virtually certain. The only reason that I don't say absolutely certain is that Tyler had seen the bird alone and it was a species he was very intent on finding. I couldn't entirely rule out that even a great birder like Tyler could have deceived himself into seeing the bird in the same way a deer hunter will turn a human into a deer as the excitement of the hunt overtakes him. I had birded with Tyler a lot, though, and he had never hallucinated a bird before. It was infinitely more likely that an Ivory-billed Woodpecker had flown in front of Tyler than Tyler had seen field marks on a bird that didn't exist.[23]

The possible existence of Ivory-billed Woodpeckers is a huge debate within birding, with some birders absolutely convinced they are extinct, and that anyone who claims to see them is clearly mistaken. Others, especially those who have reported such sightings, are perturbed at the scientific community's demand for clear photographic proof of a notoriously shy bird, claiming that the video evidence and multiple sightings should be proof enough. Whether the Ivory-billed Woodpecker is still alive or not, we can see evidence of the attempts to avoid wishful thinking even in the most wishful of searches. Even when looking for Ivory-billed Woodpeckers, called the "Elvis Bird" since they are presumed to be as rare as seeing Elvis alive, birders tried to maintain their naturalist gaze, watching the bird as objectively as possible, recording the field marks, and maintaining the institutionalized emotion of the naturalist gaze. The naturalist gaze helps birders maintain accuracy and objectivity in their observations, which, in turn, helps conservation efforts and citizen science projects.

The naturalist gaze connects the minute and the grand: the lessons learned on bird walks help birders develop a way of looking at the natural world that sees birds and people as living in a shared ecosystem. The naturalist gaze is both informed and informing, evaluative and integrative, instructive and pleasurable. I examine each of these elements of the naturalist gaze in more detail in the chapters that follow. The naturalist gaze teaches birders to appreciate common birds, helps birders evaluate and manage habitat and invasive species, and encourages birders to contribute to environmental conservation efforts through citizen science projects and other forms of advocacy.

3 COMMON BIRDS AND THE SOCIAL CONSTRUCTION OF NATURE

BIRD WALKS PROVIDE an opportunity to watch birds with other birders and to further develop the naturalist gaze. Bird walks typically consist of a leader and a group of participants who walk around a designated area and keep a list of the birds they see. Sometimes participants say they have a "goal bird," or a bird they are hoping to see. Often a bird is designated the "bird of the day," or the most exciting and rare find of the walk. However, what most often populates these walks are the common birds: the most populous, everyday birds that live in the area and that people see all the time. Unlike competitive birders who may seek to expand their "life lists" by pursuing birds they've not yet seen, the birders I studied take the time to appreciate myriad aspects of these common birds. In this chapter, I explore how birders connect the naturalist gaze to environmental issues by observing and appreciating common birds.

What does it mean for a bird to be "common"? The word "common" can refer to the name people typically call a bird: its "common name," as opposed to the bird's scientific binomial nomenclature name, such as Cedar Waxwing (scientific name *Bombycilla cedrorum*). Sometimes, the word "common" appears in a bird's common name, like Common Loon (scientific name *Gavia immer*). In this sense, "common" means the typical or expected type of loon, and such names came about when the American Ornithologists' Union created standardized lists of names. More often, when birders talk about "common birds," they are referring to any bird that can readily be found in a particular area, whether "common" is part of that bird's name or not. The American Robin (scientific name *Turdus migratorius*), for example, is one such common bird, even though it's not called the "Common Robin." It is primarily this latter sense to which I refer when discussing common birds in this chapter.

When people think of competitive birding, they don't often think about common birds. But common birds provide the largest number of species for any birder, competitive or not. Record-breaking big-year birder Neil Hayward describes the American Birding Association's categorization system as he depicts his chase for "rarities" on his big year. While he focuses on how many rarities a birder must find in order to break 700 species, he bases these numbers on the fact that a birder must see every single one of the common birds available in the United States to achieve that goal:

> Rarities are the key to a Big Year. Without them, the numbers don't add up. The ABA assigns a code to each bird on its checklist to indicate how common it is. . . . Codes 1 and 2 are for regularly occurring species (1 is common; 2 is less common). Code 3 is for rarities, such as the Rufous-capped Warbler and the Black Noddy, that are typically seen somewhere every year in the ABA region. Code 4 is for less-than-annual rarities, like the Red-flanked Bluetail and Fieldfare, and Code 5 is for the true mega rarities: birds recorded less than five times ever. I'd only seen one Code 5 this year, the Nutting's Flycatcher near Lake Havasu, Arizona, back in January. There are 666 Code 1 and 2 birds, which means that if you want to see more than seven hundred species, you need at least thirty-four rarities.[1]

The American Birding Association codes describe how common or rare a bird is for the entire North American continent. Localized birding checklists, such as those you might pick up at a local, state, or national park, describe birds' commonness or rarity relative to the particular area or region, and they make this distinction according to the season. For example, the checklist for New York City's Van Cortlandt Park contains four columns: winter, spring, summer, and fall. Each bird species that is typically, or even sometimes, found in the park gets a row on the checklist, and the birds are each labeled according to the legend:

A Abundant; seen every day in proper habitat, in numbers
C Common; seen every day in proper habitat
U Uncommon; present, not seen every day
R Rare; usually no more than one bird or small flock seen per season and not every year
X Accidental, far from regular range
E Escapee or release
* Recorded nesting
H Historical, has not nested since 1954 or before
° Non-native[2]

Birders don't only—or even primarily—spend time chasing rarities. They spend most of their time seeing the same species of birds over and over again, interspersed with some uncommon birds for their areas. Almost all of the birds seen on any given guided bird walk will include what the Van Cortlandt Park checklist categorized as "abundant" or "common" birds. Even though both American Robins and Eastern Screech Owls are technically considered "common" in much of the New York metropolitan area, people delight in seeing Screech Owls but sometimes tire of seeing the American Robins that populate their backyards every day. The birders I studied developed, and encouraged others to develop, an appreciation for even the most common birds in their area—the birds they can see every single day, such as American Robins, Red-winged Blackbirds, Rock Pigeons, Mourning Doves, House Sparrows, European Starlings, Grey Catbirds, Northern Mockingbirds, Common Grackles, Blue Jays, Northern Cardinals, and so on. Walk leaders provide taxonomic information about common birds and teach birders how to identify them, and birders appreciate common birds as individuals and not only as populations. These practices further help develop birders' naturalist gaze, and birders' appreciation of common birds has important consequences for the environment. Birders joke about common birds as a way of making fun of "listing," thus emphasizing their more holistic birding practices. Birders appreciate the commonness of common birds, since their ubiquity indicates a healthy ecosystem. Thus, common birds become a crucial medium for getting people to care about the environment—both their immediate environment and the larger ecosystem in which birds and humans all live.

ORIGIN STORIES

Learning about common birds helps birders develop their naturalist gaze. By providing taxonomic histories of common birds, walk leaders encourage participants to take the time to learn more about them. Instead of taking common birds for granted, walk leaders show what makes common birds extraordinary. The environmental historian William Cronon wrote, "By seeing the otherness in that which is the most unfamiliar, we can learn to see it too in that which at first seemed merely ordinary."[3] These taxonomic histories make common, "merely ordinary" birds seem less familiar and more exciting.

Birders defy the dualism between ordinary and extraordinary. They admire the small, the plain, the humble. They appreciate the mundane by observing the everyday, common birds with as much rigor and excitement as rare and uncommon birds. Bird walk leaders actively encourage participants to recognize the everyday beauty and wonder in the most common of birds. One way walk leaders accomplish this is to share the birds' "origin stories" of why we call birds by particular names, or how and why nonnative species of birds came to be in the area.[4] By sharing these stories, bird walk leaders make common birds extraordinary, and even the most common and abundant species of birds produce a sense of wonder.

The House Sparrow is one of the most ubiquitous bird species in any area with buildings, and *Smithsonian* magazine claims it's the most common bird in the world.[5] These birds are everywhere. And because they are everywhere, people tend to ignore them. Thus, it surprised me when, on one of the very first guided bird walks I attended, the walk leader gave the group a fairly in-depth lesson on House Sparrows. For this walk, we always meet at the park's Nature Center, a small green building in a spot where a large field meets a forest. Surrounded by trees on one side and a grassy field on the other, we often begin each walk by scanning the trees and fields for any interesting birds before starting our walk. On this day, instead of looking for other birds, Bill, the walk leader, pointed out the House Sparrows that always perched along the rooftop of the Nature Center. Bill explained, "They were called House Sparrows because they live on houses, and English Sparrows because they came over here from England on boats. But people think that originally they came to England from Africa on boats, so you could also call them 'African Sparrows.'"

I've since learned that Bill has a dry sense of humor and that "African Sparrows" don't exist, so he was likely making a taxonomic joke with that last point. Bill's willingness to spend time explaining the most common of birds on a bird walk demonstrates how walk leaders incorporate common birds into bird walks to create wonder and excitement around these birds. Even the much-maligned pigeon receives the same treatment on walks, where Bill pointed out the birds and called them by their common name, Rock Pigeons. He explained, "They're called Rock Pigeons because they like rocks. So they end up in a lot of cities because the buildings are like rocks for them."

These origin stories of common birds' common names (as opposed to scientific names) explain their history, identification markings, and behavior. On another walk with Bill, when we saw our first American Robins of the day, he explained, "They're actually a different type of bird from the European Robins found in England, but we call them Robins because they both lay blue eggs and they both have red chests and they both walk the same—three steps and then they stop." Often origin stories compare British and American names for birds, since the differences began when British colonizers came to a new continent and used the same names they used for completely different birds from another continent.[6] Luis, another walk leader, often explains to walk participants how to identify a variety of blackbirds, such as Starlings and Grackles. On one walk, Luis explained that the black birds we see are actually a completely different species than the Common Blackbird from Europe. Luis said, "Everyone knows the 'four and twenty blackbirds baked in a pie' rhyme, and we know there were Blackbirds in Europe. But when Europeans came to the U.S., they didn't have a good taxonomic system and didn't realize they were actually different birds. So, they just called them 'black birds,' even though Icterids are New World birds."

Since part of the latent purpose of sharing these origin stories is to develop an appreciation for common birds as part of the naturalist gaze, walk leaders correct themselves if they accidentally take common birds for granted. Justin, another walk leader, did so when we came across two common birds, the American Robin and the Mourning Dove, on a walk. Justin stopped, pointed out a bird, and said, "Oh, it's just a robin." He immediately corrected himself: "Wait, we shouldn't say 'just a robin.' Robins are perfectly nice birds." A few moments later, Justin pointed out a Mourning Dove, and exclaimed, "First Mourning Dove!" Justin was so excited that one of the participants asked, "Is that your first time ever seeing a Mourning Dove?" Justin replied, "No, it's the first Mourning Dove of the trip! We should be excited!" Walk leaders attempted to instill an appreciation for common birds through origin stories, and they tried to maintain that appreciation in all aspects of their walks.

IDENTIFYING COMMON BIRDS

One of the primary ways that walk leaders encourage participants to appreciate common birds is by teaching them how to identify, differentiate, and distinguish among common birds. These birding skills further develop birders' naturalist gaze. Walk leaders teach participants how to observe common birds in a holistic and appreciative manner, which contributes to the instructive element of the naturalist gaze. Walk leaders also help new birders learn how to identify birds using field guides and scientific research, which contributes to the informed element of the naturalist gaze. These elements of the naturalist gaze intersect when walk leaders teach participants how to identify and appreciate common birds. In chapter 1, I showed how walk leaders encouraged new birders to use common birds' size as a way of describing birds for identification, as a very first step in learning to identify any birds. In this section, I show how walk leaders teach birders how to properly identify common birds.

Walk leaders take the time to teach participants about birds that many people likely know how to identify, such as American Robins. On a walk with Justin, one participant pointed out a robin to be added to the day's species list, noting, "He's a big, fat robin. Very pretty." Justin agreed and added, "That's right—we shouldn't take them for granted." Diane, another walk leader, encourages birders to use common bird identification to improve their birding skills. Diane emphasizes the importance of birding by ear, and she always makes a point of identifying a robin's song: "I wanted to point out the robin's song, even though it's common, so people can recognize more bird-song." Diane acknowledges that American Robins are common, but she doesn't dismiss their importance, since she points them out on every walk.

In the volume *Trash Animals: How We Live with Nature's Filthy, Invasive, and Unwanted Species,* the American Studies professor Charles Mitchell notes that fully half of the Federal Aviation Administration's reports on bird strikes with airplanes don't specify the species of common birds involved in a strike—instead, they refer only to "blackbirds."[7] In contrast to this generic, though technically correct, term (the birds *were* all black), walk leaders help participants appreciate common birds by teaching them to differentiate among common birds that could easily be mixed up, such as the numerous varieties of black birds. Instead of calling them "blackbirds," walk leaders teach participants the differences among European Starlings, Common Grackles, Brown-headed Cowbirds, and Red-winged Blackbirds, all common blackbirds.

While waiting to begin our walk one morning, all of the participants engaged in the typical prewalk practice of scanning the neighboring field and trees for any interesting birds. A large group of European Starlings and Common Grackles milled about on the ground right next to us. While more competitive birders may have ignored the birds or simply checked them off the day's species list, this group had been participating in walks long enough to have developed a naturalist gaze. Instead of ignoring them, one participant asked Justin, the leader, "Can you explain the difference between a Starling and a Grackle?" Justin eagerly responded by pointing out the birds on the ground: "The juvenile Starlings are all brown, and the adult Starlings are black with a yellow beak. You can tell the difference since Grackles have a black beak, but Starlings have a yellow beak. Starlings also have a very short tail."

Even though many birders do not like Brown-headed Cowbirds because of their parasitic nature, walk leaders still teach participants how to identify and distinguish them from other blackbirds.[8] As we walked through a field with Bill, he asked the group to stop so he could look at a large flock of birds through his binoculars. "Those are Grackles and Red-winged Blackbirds, but I'm hoping to see a Cowbird," he explained. He started to walk off and then stopped. "Yes! There's a Cowbird. You can tell by the way they walk. That's what tipped me off." A walk participant who was keeping the species list said, "That's a new one for today!" and added Brown-headed Cowbird to the list.

FIGURE 3. Red-winged Blackbird (photograph courtesy of Dave Saunders).

A final type of blackbird participants learned to identify was Red-winged Blackbirds, distinguished primarily by the red bands on their shoulders and by their distinctive song. As the Cornell Lab of Ornithology describes them, "Male Red-winged Blackbirds do everything they can to get noticed, sitting on high perches and belting out their *conk-la-ree!* song all day long."[9] Despite the commonness of the Red-winged Blackbird, walk leaders always take the time to describe these birds and their distinguishing characteristics. You've almost certainly heard a Red-winged Blackbird, even if you didn't realize it. Their loud song pierces through many a birding walk, and instead of ignoring it, walk leaders take the time to identify and explain it. The Red-winged Blackbird's song was one of the first I learned as a novice birder, precisely because of the attention that walk leaders gave to it. On a walk with Bill, he pointed out the call and explained, "They call for two main reasons: one is to attract a mate, and the other is to mark their territory, to say, 'Hey buddy, this is my spot, stay away.'" Birders appreciate their looks, too, as Tom did when we spotted one on a walk: "For such a common bird, when Red-winged Blackbirds spread their wings, they really look spectacular."

Walk leaders also teach about the differences between male and female Red-winged Blackbirds. Like most species, Bill explained, "the males always have the bright colors, and the females don't. But if both males and females of a species are brightly colored, this means they both take care of the babies."

Female Red-winged Blackbirds are particularly drab. Despite her status as a "little brown job"—a brown, nondescript bird that is difficult to identify because of a lack of distinguishing characteristics—this common bird can become the bird of the year when she is out of range. When a female Red-winged Blackbird showed up in an outlying Scottish island, British birders immediately went to see her, by driving, taking ferries, and even chartering planes. What is common in North America is rare elsewhere, and this was the first-ever sighting of a Red-winged Blackbird in Europe. Thus, despite her drab exterior, her mere presence in Scotland caused a stir. As one of the twitchers who made the trek to see her said, "The male is quite smart, but the female is as dull as anything. Still, it doesn't resemble any of our birds and that makes it pretty special."[10] As I discuss further below, one person's common bird is another's bird of the day. Even though it is a common bird, walk leaders encouraged participants to understand and appreciate the Red-winged Blackbird. By taking the time to distinguish, identify, and appreciate the most common birds that they see every day on walks and in their yards, birders further develop their naturalist gaze.

APPRECIATING INDIVIDUAL BIRDS

Environmental sociologists tend to view animals as populations, not as individuals.[11] This is especially true for farmed animals and for wildlife.[12] With the rise of human-animal studies, more scholars see animals as individuals, with a self and a sense of agency.[13] Watching wildlife through the naturalist gaze means that birders learn to appreciate common birds both as species and as individuals. The close observation of individual birds helps develop this element of the naturalist gaze.

The academic divide over animals as individuals mirrors disagreements between animal rights advocates and conservation proponents. In the late 1990s, a debate in Wisconsin grew over whether to allow the hunting of Mourning Doves. The two groups against the hunt brought together animal rights advocates, who opposed all hunting, and conservation advocates, who supported hunting but opposed this particular hunt. The sociologists Ann Herda-Rapp and Karen Marotz argued that only the animal rights advocates appreciated individual animals, and that the pro-hunting but anti–dove hunting group focused on the need to conserve the population of doves.[14]

I heard this same argument on my visit to the National Audubon Convention, when I explained to a birder that my previous research was on animal rights activists and my new research project was on birders. This birder explained to me that one major element that I would find different between the two groups was that animal rights activists care about individual animals, and birders care about populations. However, I found in my research that birders do appreciate individual animals, even individuals of the most common species of birds. In our interviews, birders told me they appreciate individual birds, even pigeons, as Ellie explained to me. She said that watching individual birds "can be really interesting to look at the behaviors," but she also likes to identify and keep track of the birds she watches, in contrast to her boyfriend, who is not a birder: "My boyfriend is like, 'Why do you want to identify?' He doesn't care if he puts a name on it, he's more interested in just watching something do things. And I'm like, 'That is what we do. Otherwise, why would we want to see the same birds over and over and over again?' I love to watch pigeons, actually. At train stations and things like that. And they can be beautiful, the plumage."

Another academic misconception of birders is that they automatically dismiss "little brown jobs" or that they view common birds as "garbage" or "trash" birds.[15] My findings counter such claims. In our interview, Tom cautioned against dismissing any kind of bird because the bird may be considered common. He said, "I don't think there is such a thing as a common bird. They all have their own personality, they all have something special about them." In addition to valuing individual European Starlings' personalities, he also delighted in flocks of starlings flying in murmurations: "People look at starlings and say, 'Oh yuck.' If you ever watch how they do this flight thing called murmuration, I don't know if you've ever seen it, where they fly, and it looks like a school of fish in the air, where they're just going like this and that. It's an absolute ballet happening in the air right in front of you. Sometimes I've seen thousands of them doing this at once. In the evening, at [the marsh], you watch them coming in to roost for the night, and they come in, ten thousand birds come in, and they're flying in flocks of a thousand, two thousand, and they're doing this thing, it's like a ballet." Tom's example of why starlings are special relies on their working together in flocks, but he still argues that individual birds deserve our appreciation, as when he later critiqued the term "junk bird" and said, "Every bird's got something special about it."

Likewise, when I interviewed Catherine and Lenny, they described how their close observations of individual birds for a Breeding Bird Atlas helped them appreciate individual common birds "that you kind of take for granted because you see them all the time," as Catherine put it. She continued to describe her interest in birds' entire life cycle "that you may not necessarily notice when you're just walking and you see, 'Oh look! There's a nest there!' And then it becomes something really cool." Lenny added, "It's surprising how much isn't known about a lot of the common stuff," and then Catherine shared this story about their neighbors: "It's kind of cool when you uncover where their nest is. Our next door neighbors who aren't really nature-oriented at all, we found a nest in one of the trees in their yard. So I was standing there one day, watching the bird bringing food in to the babies, and she kind of looked out of her door and goes, 'What are you doing?' I said, 'Well, you've got a, there's a robin's nest in this tree.' And she goes, 'Is that a good thing?' And I said, 'Absolutely, any time nature can accommodate, or override what we've done to the habitat, it's a great thing.'" Catherine and Lenny took the time to share their interest with their neighbor, who wasn't aware that birds were nesting in her tree. They explained that watching each individual bird's behaviors led to a deeper appreciation of birds as individuals. In their research on a museum exhibition of photographs of wild animals, the sociologist Linda Kalof and her colleagues found that viewing photographs of individual animals encouraged viewers to see animals as individuals, and not simply as generic members of a species.[16] They argue that the opportunities to view individual animals in museums, and even in conservation exhibits in zoos, offer important chances for people to understand animals as individuals. While museums and zoos may offer such an opportunity, I argue that watching wildlife through the naturalist gaze provides an even better opportunity to see wild animals as individuals, since the wild animals can interact with their natural habitat and with other animals.

Even when birders weren't watching individual birds' behaviors, they still highlighted the importance of appreciating birds as individuals. On one bird walk in North Park, a participant noted a Cardinal's song. "I can see Cardinals in my own backyard," someone scoffed. "But they wouldn't be these particular Cardinals," another person added. "Yeah, these are *North Park* Cardinals!" said a third. Even though Cardinals are so common that they are found in nearly everyone's backyards in this area, the participants realized that the Cardinals they saw on this walk were different individuals than the ones in each of their own backyards, and thus they warranted attention and appreciation. In these many ways, birders develop and share an appreciation for common birds as individuals.

DEVELOPING AN ECO-SELF

As mentioned earlier, academic researchers sometimes misconceive of birders as maligning common birds as "trash" or "garbage."[17] In contrast, I've shown how birders appreciate common birds both as species and as individuals. Perhaps these academic misconceptions come from taking birders' lighthearted disparagement of common birds at face value rather than placing it in a broader context, such as the naturalist gaze. Birders use humor to make fun of themselves as birders, with common birds as a foil. This humor comes from the most devoted and conservation-minded birders, who jokingly disparage their own need for common birds when compiling their species lists. The birders I studied contrasted their holistic focus on birding with a more competitive approach to birding that's purely focused on the number of species on one's list. Such a holistic practice represents an outcome of a well-developed naturalist gaze. These birders turn the naturalist gaze back on themselves and the practice of birding itself, and in doing so, they develop their eco-self.

The environmental sociologist Stella Čapek described how people develop an eco-self, or an ecological identity, in her study of the destruction of a Cattle Egret nesting colony.[18] In 1998, a crew of workers extending a road for a new subdivision bulldozed over a large portion of a Cattle Egret nesting colony, killing or seriously injuring close to 5,000 birds. The surviving egrets moved into surrounding neighborhoods, and residents in the area had hundreds of Cattle Egrets in their yards, many of them injured or dying. Local residents reacted to the birds in different ways. Some saw the birds as pollution and wanted to get rid of them so that the landowners could develop the property. Others saw the birds as innocent, wild, and beautiful, calling "greedy" humans the real problem. Still others changed their view of the birds from negative to positive through their participation in the rescue effort. In doing so, they developed what Čapek called an "eco-self." The eco-self describes our part of the self that is connected to nature. Čapek's work helps us understand how people's sense of nature and wildlife may change, but in her work, the eco-self is mobilized only through a horrible incident. Birders develop an eco-self without the need for any kind of horrible incident or "moral shock" to get them to view nature in a sympathetic or symbiotic way.[19] Birders develop their eco-self in a variety of ways; in this section I focus on their use of humor when discussing common birds, to demonstrate how common birds play a central role in birders' naturalist gaze.

Sometimes bird walk leaders and participants use humor when discussing common birds, indicating the commonness of these birds. Such humor often revolves around species lists, or the lists that the leaders keep of all the birds seen on a trip. For example, just as we started one walk, the leader said, "Oh! There's a House Sparrow. Let's go ahead and check it off the list!" and laughed. Participants sometimes get in on this humor as well, when they point out birds to the leaders. On one walk, as the group tried in vain to find any birds in a marsh, everyone searched intently but found nothing. Suddenly, we saw a Red-winged Blackbird, and a participant exclaimed, "Oho! A Red-winged Blackbird!" in a sarcastic tone, as if to make fun of their collective excitement over seeing such a common bird. Or, when a participant approached the group just after a robin flew away, the leader let out a low whistle and chided, "That's a big miss—I don't know if we'll get that one back," and everyone laughed.

Some days, field trips don't generate many species for the list. A good day garners twenty to thirty species, a great day thirty to forty species, but on other days, the group is lucky to see a dozen species of birds.[20] Some days, the group sees only common birds. On such days, the walk leaders often apologize to the group for having seen only common birds. At the end of one such walk, on a foggy morning with poor visibility, we ended the trip with very few species. As we walked back to the parking lot, everyone thanked the trip leader. He said, "I wish it were a more exciting trip," but when we reminded him about the Scarlet Tanager we all enjoyed seeing, he conceded, "You're right, that was a good sighting."

In the field, walk leaders sometimes acknowledge a bit of disappointment when what they thought was a rare bird turns out to be a common bird. On a walk with Diane, she saw some movement and exclaimed "Whoa! Whoa!" and then realized, out loud, "Oh, it's just a chickadee." Even walk participants expressed this kind of frustration. In one instance, the group stopped to look at an Eastern Towhee, and we spent a long time looking through the scope, taking photos with zoom lenses, and generally admiring this bird. One man, who was ready to move on, joked that this reminded him of spending twenty minutes on a walk looking at chickadees. "It's a chickadee," he said. "Let's move on."

While birders lampoon common birds for their commonness but still admire them as individuals, birders also admit that they appreciate common birds because they're important for their species lists. On walks with groups of more experienced birders, such as the field trip I attended with the Audubon Council of New York State, which brought together Audubon leaders from around the entire state, the group did not always stop to observe each and every bird. Instead, these experienced birders called out the species names as they saw or heard common birds, such as Eastern Mockingbirds, Catbirds, American Robins, and the like. Their names went on the species list, and we continued the walk, on the search for rarer birds.

On a regular weekend walk with a local Audubon chapter, the group watched Wood Ducks on a pond for about fifteen minutes, during which time the walk leader and participants pointed out the Wood Duck boxes and each individual Wood Duck that we saw, but everyone ignored the dozens of Canada Geese that were flying all around the pond and generally making a racket. These geese all made it on to the walk's species list, but the walk leaders sometimes jokingly dismissed them. When a participant pointed at the sky and told Diane, the leader, "Hey, there's three Canada Geese to add to your list," Diane deadpanned, "Okay," without even looking up, and everyone laughed. Diane's dismissal of the Canada Geese was performative, for humor, and facetious, since she spends most of the time when she leads walks teaching participants about common birds.

If birders truly appreciate common birds, why would they act like this? Rather than seriously maligning common birds, these birders' jokes are lighthearted and rhetorically designed to emphasize common birds' importance to birders' purported bottom line—the precious species lists. They are, therefore, making fun of birders. They distinguish their own eco-selves as conservation-minded birders, as opposed to a stereotype of "listers," who merely count birds and move on, seeking only to expand their species lists. Chloe said: "For me, it's not just about 'check, done, saw it, move on.' For me, it's about the ability to just see them. So it's just a robin, fine. But what is that robin doing? Is it, what is his behavior like? Is it carrying food? Is it trying to pull that worm out of the ground? They're so silly, to see robins do that. Even the most common birds have amazing things to see."

While the conservation-minded birders I interviewed and observed appreciate birds as individuals, they also make fun of their own dependence on common birds for creating robust species lists. In our interview, Rhonda told me that she primarily values common birds for their ability to add numbers to a species list: "I know when we go on walks, you want to be able to identify all the birds. And you want to get as many species as you can. Robins, starlings, pigeons, you don't really go out looking for them, but you really do end up wanting them because it adds to your list. It makes your list longer."

Walk leader Diane joked about this as well when we added common birds to the species list. On a walk, she said in a sarcastically excited tone, "Look, a Rock Dove! AKA pigeon! Just trying to pad the species list." When I met Marcus at the World Series of Birding, he proposed that species lists create a sort of equality among birds in such competitions: "Actually, something like the World Series of Birding, you need the common birds. You start with the common birds. They all count equally. That's kind of a neat concept of the World Series of Birding." The practice of listing species puts common birds on the same playing field as rare or more exciting birds, in the sense that each species counts equally toward the group's species list. You don't earn more points for finding a rare bird; a common bird counts just as much toward the group's final species list number. In this way, common birds play an even more important role for competitive birders engaging in a big day competition. For the conservation-minded birders I interviewed and observed, common birds also simply count as individual animals deserving of our attention and admiration.

Other naturalist groups don't seem to exhibit this same naturalist gaze or eco-self. Of mushroomers, the cultural sociologist Gary Alan Fine writes: "Mushroomers feel so strongly that they discourage colleagues from wasting time trying to identify small, dingy mushrooms. . . . 'We don't have the books or chemicals to identify them. It's not worth the time.'"[21] To birders, it *is* worth the time to identify birds, and they readily spend time in the field consulting field guides, online images and sounds, and one another to make the identification. Perhaps part of the difference between birders and mushroomers, in this example, comes from the fact that birders can identify birds without relying on chemical analyses or other high-tech solutions.

Underneath all of this joking about common birds, the birders I interviewed and observed showed they really did care about common birds.

Shortly into our walk one day, in a park in New York City, Justin, the walk leader, said, "I want to make sure we keep track of all the birds we've seen. We've seen almost everything but a pigeon. Where's a pigeon when you need one?" Everyone laughed, and then someone pointed out a robin. Justin then earnestly shared, "Robins are perfectly beautiful birds. Just because they're common, that doesn't mean we should ignore them. Every day is an opportunity to see a new pigeon, or a new robin. We should get excited about that." In these ways, birders showed how their eco-self appreciated common birds, in contrast to listers, who may seek only to check a bird off the species list and then move on to the next one.

THE SOCIAL CONSTRUCTION OF COMMON BIRDS

Up to this point, I have shown how guided bird walks teach new birders to appreciate common birds. New birders learn origin stories and taxonomic information about common birds, how to identify common birds, and to appreciate common birds as individuals. These lessons in developing their naturalist gaze affect birders' sense of self as birders, as seen in the previous section on their eco-self. Another effect of the naturalist gaze is that birders appreciate commonness in and of itself. They learn that the presence of common birds indicates a healthy ecosystem, and thus common birds become much more important than they may seem to be at first. By viewing common birds through the lens of the naturalist gaze, birders learn to view the natural world in a new way. Environmental sociologists call the development of a new perspective on the natural world the social construction of nature.

New birders learn to take the time to watch common birds, but they must continue to develop their naturalist gaze and their interest in wildlife conservation to better understand why common birds matter. More advanced birders, with a well-developed naturalist gaze, view common birds as important elements of our shared ecosystem. They learn to see the exotic in the common, rather than ignoring it. One group elaborated this perfectly when we took a bird walk with a couple preparing to move to Florida. A pair of Great Blue Herons flew over, and the couple said, "Those are going to be our yard birds," meaning the common birds they would see every day in their yard. "We'll have so many warblers in our yard in Florida, we might get sick of them!" they continued. (Warblers are highly sought-after songbirds that most birders in the New York metropolitan area get to see only during spring and fall migration.) Rhonda, the local Audubon chapter president, responded, "Listen—a Starling is on someone's life list. A Canada Goose is on someone's life list." Rhonda turned the concept of a life list on its head by pointing out that the birds we may perceive as common are exotic to someone else, and that we should appreciate all birds equally. While someone might cynically interpret this as indicating that "life lists" still primarily motivate birding goals and desires, Rhonda's comment gives common birds more value than they typically receive. Her comment also demonstrates the variety of ways that birders can view the exact same birds—Great Blue Herons and Starlings alike can be seen as "common" yard birds, and a different person can see them as exciting birds to add to their life list. The ability to hold differing views of the exact same natural phenomenon, based on one's location in society or in the natural world, is what sociologists mean by the social construction of nature.

In their work on landscapes, the rural sociologists Thomas Greider and Lorraine Garkovich argue that any one natural phenomenon, like an open field, may mean different things to different people, such as a farmer, a hunter, or a real estate developer. The exact same physical entity of an open field is symbolically transformed to reflect each person's definition of what the field means to them. They write: "These symbolic meanings and definitions are sociocultural phenomena, not physical phenomena, and they transform the open field into a symbolic landscape."[22] People give meaning to nature and the environment through the filter of their own values and beliefs. How we view the environment is grounded in our cultural belief systems. Not only do we define nature, but Greider and Garkovich argue that our understandings of nature, and our relationships with the environment, help create our own definitions of self—they help us understand who we are. Birders socially construct common birds to be valuable elements of a shared ecosystem as they develop their naturalist gaze, their eco-self, and their approach to wildlife conservation.[23]

Sociologists seek to "make the familiar strange," to try to explain phenomena close to home with the same scrutiny that we use to inspect cultures or practices that may be different from our own. Similarly, the presence of other birders, from other flyways and other ecosystems, helps remind birders to do the same with their own backyard birds—to see them in a new light. Amy told me that when her cousin came to visit from Canada, her

FIGURE 4. Blue Jay (photograph courtesy of Dave Saunders).

reaction to our common birds helped Amy realize that the birds she previously considered common were actually quite interesting and worthy of attention: "It takes someone coming in from outside to kind of remind you of that. This cousin was here from Canada, and she was just raving about our robins and our cardinals, because they don't have them up there. And she was, 'Oh, look at the color on that! Isn't that beautiful!' And I just said you're right, we take—oh, it's just a robin, or it's just a cardinal—we take it for granted." Visitors like Amy's cousin, or birders' own travels to other countries, help birders further appreciate common birds and develop their naturalist gaze. Birders learn to evaluate the common birds in their ecosystem as valuable.

Many birders travel to go birding because they want to see particular birds, expand their life lists, go hiking in new areas, and generally appreciate the natural world in as many places as possible. This travel, combined with encountering visitors to their local areas, reminds birders that their common birds merit appreciation, as Vivian noted: "The cool thing about when you travel, too, you realize that there's people who are coming here to see our everyday—our regular birds are exotic to them. Like a Blue Jay is an amazing bird. That kind of thing. People would come here and be amazed by that, and it flies by us, and we're like, 'Eh, whatever.' So yeah, you try and take the time to really look at a bird, or if they're doing an interesting behavior that maybe you never noticed before."

Many birders shared this social construction of common birds: a bird that is common for one may be exotic to another person. Why are common birds important? On bird walks, sometimes it seems like the walk leaders take the time to appreciate common birds because they have no other choice—common birds may be the only birds the group sees. The desire to see any birds on a walk does not explain why birders value common birds. The primary reason that birders appreciate common birds is because of their commonness. The presence of common birds means that an ecosystem is thriving. Common birds serve as indicator species, or indicators of ecosystem health. An indicator species refers to a species of plant or animal whose presence, absence, or general well-being indicates the health of the ecosystem in which they live.

Indicator species provide a method of indirectly measuring ecosystem health, as Ethan explained in our interview at the World Series of Birding: "These birds are common for a specific reason, and they are indicators, in my opinion, and there's a lot of scientific basis for why a particular species is very abundant, and how they can indicate the quality of their environment. If I started seeing a decline in American Robins, I would notice that something is probably different, or something has changed in our environment, and that is a sign for concern. And so I don't think anybody should just disregard robins or something like that, tune them out." At eighteen years of age, Ethan was, by far, the youngest birder I interviewed for this project. Even at his relatively young age, Ethan had already developed his naturalist gaze such that he appreciated American Robins for the ecological information they provide. He went on to say that birders need to ask themselves if they are seeing or hearing the same amount of robins that they normally would, and then go on to ask, "Is this a seasonal change, or is this due to something else? Are other environmental factors, are humans causing it? Or is it natural processes?" These questions are best answered by common birds, rather than rare birds, as he explained, "Common species give you a whole different array of questions that you can pose in your mind, because you know so much about them." It is precisely American Robins' commonness that allows birders to use them as environmental indicators. Robins' commonness takes on a new, and much more important, meaning when viewed from this perspective.

"Keeping common birds common" serves as a slogan for the National Audubon Society, BirdLife International, and many other birding organizations. Birders use the extinction of the Passenger Pigeon as an example of what could happen if people ignore the health of common birds, as Kristina explained to me when I interviewed her at the National Audubon Convention:

> One of our themes is "keeping common birds common" and we tell the story of the Passenger Pigeon, and the extinction of the Passenger Pigeon, who at one point was four to six billion in numbers, and then down to zero 100 years ago, and so one of the things I say to kids all the time, is back when John James Audubon was alive, they never realized that birds could go endangered or extinct. It looked to them like there was an infinite number of birds. And I say to them, "Think about now, if I step on an ant hill. Do you think that I've killed most or all of the ants in the world?" And they're like, "No!" Well, that's how they felt about birds 200 years ago. But now we have the tools and resources to know that that isn't true, and so we have to use those tools and resources to make sure that common birds stay common.

Kristina went on to warn against people ignoring common birds and saying "'ho hum,' because they see them all the time." We now understand that species can go extinct, which means we must watch out for even the most common of species. Kristina's comments also demonstrate how the naturalist gaze is historically bound—in the nineteenth century, wildlife observers did not fully understand the causes of extinction. The Passenger Pigeon was so common that observers never considered its possible demise. Both amateur and professional hunters killed Passenger Pigeons for food, and they disrupted the birds' nesting grounds, which led to a fairly quick path to extinction from the late 1800s to the early twentieth century, when the last Passenger Pigeon died in captivity in 1914.

The plight of the Passenger Pigeon also exemplifies problems with how people conceive of, or socially construct, nature to be entirely separate from our human societies. The environmental historian William Cronon described it as "a dualistic vision in which the human is entirely outside of the natural."[24] People have created a dualism, a symbolic divide, between human society and nature. This divide further distances people from nature, especially from the nature found in our own backyards. The more people view nature as something "out there," the less we view nature as something that we can find in our everyday lives. Cronon's primary critique of this divide is that it makes people ignore the beauty in the everyday nature that is all around them: "My principal objection to wilderness is that it may teach us to be dismissive or even contemptuous of such humble places and experiences."[25] Instead of dismissing the "humble," small, or plain, birders celebrate it by appreciating common birds for all they embody—their beauty, their environmental benefits, and their fragility.[26] The knowledge that even the most common of birds can go extinct helps remind Dawn to appreciate common birds, as she explained in our interview:

> I think you can watch something common like a pigeon or a robin, and every time you watch that individual or individuals, they're going to have different behaviors, they're going to be doing different things, there's subtle differences between each one, even though they're the same species, they're going to have color variations. You don't want to take for granted the common birds because you can see some pretty cool stuff if you watch them long enough or close enough. Also, sometimes common becomes no longer common. Passenger Pigeons were extremely common, and they went extinct. And some of our species that were considered common decades ago are just not as abundant, so you hate to take that for granted.

Dawn extolled the virtues of observing common birds because their commonness or abundance may be fleeting, but also because closely observing them allows the watcher to note variations in bird looks or behavior. Her comments exemplify how the naturalist gaze is pleasurable but also concerned for the health and well-being of birds in their natural habitat. Common birds also provide an entry point for understanding environmental issues.

COMMON BIRDS AND THE ENVIRONMENT

Finally, birders with a well-developed naturalist gaze relate common birds to larger environmental issues. Appreciating common birds is an important way that people learn to care about the environment, both locally and globally. By socially constructing common birds to be significant environmental indicators, birders view common birds as providing information about and connections to high-stakes phenomena such as climate change.

When the National Audubon Society was founded in 1905, six out of every ten Americans lived in a rural area.[27] With urbanization, those numbers have drastically changed, with slightly more than one out of every ten Americans living in a rural area.[28] We used to live on farms, with animals all around us. Now, the only time we see nonhuman animals is in our homes as pets or in zoos as entertainment. The art historian John Berger describes this change: "Zoos, realistic animal toys and the widespread commercial diffusion of animal imagery, all began as animals started to be withdrawn from daily life. One could suppose that such innovations were compensatory. Yet in reality the innovations themselves belonged to the same remorseless movement as was dispersing the animals."[29] As animals disappeared from our everyday lives, Berger argues that we created new, commercial ways of bringing animals back into our lives. But, he notes, those methods take precisely the same form as those that removed animals from our everyday lives in the first place.

In contrast, birders find ways of seeing animals in their everyday lives that do not negatively affect the animals. The naturalist gaze encourages birders to socially construct nature in a vaster, broader sense than our increasingly urbanized lives typically provide. Birders appreciate common birds because birds tie birders to their immediate environment—whether rural or urban—and common birds provide an entry point for understanding the natural world, as Sue said in our interview: "These are the birds that are going to incite you to learn more. Mainly because you see them every day. They're your neighbors. They're your colleagues, they're your outside the window visitors. And then sometimes, you're going to say, well, I know robins, and I know house finches, and I know—who is related to robins? Who is related to house finches? And it's going to expand your interests. So, it's critical that you learn the common birds first." Sue noted a process I described in chapter 1, where learning about common birds helps birders learn to identify other, less common birds. Sue also highlighted the notion that the birds people see every day are the ones that draw people's attention and encourage people to learn more about birds and about environmental issues related to birds. Understanding common birds helps people relate to their immediate environment, which can then help people relate to the bigger picture of the health of an ecosystem, or the entire planet, as Allen explained in our interview:

> It frustrates me that if people see a starling, they think everything's fine, because there's still birds out there. It may be a starling on a bittersweet vine, but you

> know, you've got a plant out there and you've got a bird out there and so everything's fine with the environment. And I think that's part of what we need to do, is educate people. I think birds are one of those charismatic animals that people can relate to. It's not some far-off polar bear that you'll never see or never quite understand. But people know robins and cardinals and I think if that's a tool to get people to take care of their environment better, appreciate the environment.

Allen opined that if people see an invasive bird on an invasive plant (e.g., a starling on a bittersweet vine), they think everything is fine, because we still have birds and plants. Allen wished more people would understand that invasive species do not mean that an ecosystem is thriving. His method for this is to encourage people to appreciate the most common birds in their yard: "Get them to like cardinals, then they like something else, and like something else, and pretty soon they might even like snakes and some of those nasty little animals that a lot of people don't like."

Allen's point illuminates a key issue for climate change scientists—how to get people to care about or understand issues that they think don't directly affect them. The Yale Program on Climate Change Communication surveyed Americans on their attitudes and beliefs on climate change, and it found what it calls "Six Americas," or six different ways of understanding climate change.[30] The largest group, "The Concerned," represents 33 percent of Americans. The people in this group believe climate change is happening, but they have not yet engaged personally on the issue—they think climate change is far off and won't affect them personally. Polar bears, the emblematic charismatic megafauna of climate change, remain far off to most people. Birds, in contrast, serve as another indicator species of climate change, and common birds are readily available for everyone to observe in their own backyards. Viewing common birds through the naturalist gaze can bring seemingly far-off issues like global climate change to people's everyday lives.

The National Audubon Society recognizes the connections between common birds and climate change, and it tries to capture people's environmental attention through the common birds in their own backyards. In a telephone town hall meeting that the National Audubon Society held in January 2017, Chief Network Officer David Ringer explained the importance of connecting people to birds or to special places that bring meaning to their lives. They don't have to be "exotic birds" or "faraway places," he explained:

"The best way to get people involved locally is to focus on birds that are most meaningful in daily life, and to focus on local parks and rivers, where people visit with their families." National Audubon found that the most popular birds are the Cardinal in the east and the Stellar's Jay in the west—the birds that people see every day. He added, "When people notice that birds they have known their whole lives are moving to new places because of climate change, that sends a powerful message."

Audubon wildlife educators understand the importance of this connection as well, and they try to teach children how to identify the common birds that they see most often. In our interview, Cynthia told me that her primary goal as an environmental educator is to get children to learn and appreciate the birds they see every day: "Not appreciating those birds would be an incredible disservice to my students because they might never see a Bluebird, or an Oriole. They actually think Robins are Orioles all the time, and so I try to make that distinction. We have a 'bird bingo' game where they are trying to find the different species on their board, and it's not just birds. There are some plants, or insects and things. But Pigeon is on there, and House Sparrow is on there, and I don't really get into the topic of invasive species, because they're, you know, eight. So really the point is just getting them to appreciate and notice what's around them." Cynthia teaches her students to understand the common birds, common plants, and common insects that they are most likely to see around them. Even though she knows that understanding the difference between native and nonnative species will be important for their ecological education, she does not broach that topic because of their young age. She is taking the first step to get them to appreciate the nature in their own backyards, with the hope that in the future they will continue this education. Thus, common birds provide an entry point, and an important medium, for getting people of all ages to care about the environment.

This chapter showed how birders understand nature and give it meaning through the meanings they ascribe to common birds. People tend to think of nature as something distant, but birders socially construct nature to be everywhere by paying as close attention to common birds as they do to exotic, rarer birds. New birders learn about common birds on bird walks, where they learn common birds' origin stories and identification markings. These lessons further develop new birders' naturalist gaze. Viewing common birds through the naturalist gaze affects birders' sense of self and the environment, as birders see common birds as connecting the minute and the grand—the individual birds birders see in their own backyards can carry messages about the ecosystem and climate. Once birders cultivate this appreciation of common birds through the naturalist gaze, they can develop new ways of understanding wilderness and wildness, which I discuss in the following chapter.

4 WILDERNESS, WILDNESS, AND MOBILITY

IF YOU EVER talk to a birder, you may notice that they don't always look at you when you're speaking—or even when they're speaking. Birders aren't rude; they're just looking for birds. Anytime birders are outside, their attention is easily captured by any movements or sounds in the background. In July 2015, I interviewed birders at the National Audubon Convention, in Leesburg, Virginia. The convention was held in a large conference building overlooking a golf course. Staying inside all day to talk about the importance and joy of birding drove conference attendees outside to enjoy the sunshine and fresh air whenever they got a chance, and I conducted many of my interviews outside on a deck, under a canopy of trees surrounding the golf course. During the course of these interviews, birders frequently interrupted themselves to point out a silhouette of a bird flying overhead and identify the species. Therefore, a perfectly normal conversation with a birder can go something like this, taken from my interview with Allen at the convention: "I don't know if I'd go to moths, but wolves, or grizzly bears, or some animals that people are familiar with, and you have the opportunity if you're in the right areas to see them. I'd be curious how people look at—Great Blue Heron [*points up*]—how people would feel about, whether it's black bears or white-tailed deer, how their answers would be compared to birds or something that has some more advocacy going behind it." Allen didn't skip a beat in his thought process; he simply stopped to point out the Great Blue Heron flying overhead and then went right back to making his point about wildlife conservation.

After I'd been birding about a year and had greatly improved my identification skills, I found myself doing this in conversations with my colleagues at work. Our campus sits on one hundred acres in the middle of the bustling New York City suburbs and teems with native trees, shrubs, and bodies of water, all of which attract birds. While standing outside and chatting about work, I started to find myself looking over my colleague's shoulder to watch a White-breasted Nuthatch climbing up a tree. Realizing that I had become distracted, I pointed out this Nuthatch to my colleague, and we shared a bit of enjoyment of nature before getting back to discussing work. By always looking out for birds, birders find nature and the wild wherever they are. I explore this phenomenon in this chapter.

Birds' mobility and ubiquity provide unique opportunities to view wild animals in the wild in different ways than how people typically watch wild animals in captivity. Birds' omnipresence allows people to connect to nature, rather than seeing nature as an entity separate from the areas where people live. Birds exemplify wildness, and birds' mobility brings wilderness with them wherever they fly. Birds move of their own volition in ways that trees and rocks in wilderness areas cannot.[1] Birds maintain their wildness, even in the middle of cities, by persisting as wild animals in these built environments. Birds feed, nest, breed, and fly in and out of developed areas. Birds that live in cities and wildflowers that grow in the cracks in a sidewalk in a city resemble one another in that they are both wild beings persisting outside of wilderness areas. But they differ in important ways—birds can move, but flowers cannot. Birds' omnipresence, combined with birders' practice of always birding, further develops the naturalist gaze, through which birders come to view themselves and wild animals as part of a shared ecosystem. Birders can always watch birds in this shared ecosystem, birds' mobility brings wilderness and wildness to birders, and birders create a little patch of wilderness for and with birds in their own backyards.

ALWAYS BIRDING

Since birds are everywhere, birders watch birds everywhere. Birders are always birding. This means that the naturalist gaze is always "on." The naturalist gaze does not describe a perspective that birders use only when they are on a bird walk; birders always use the naturalist gaze, since birders continually look for birds wherever they are. Chapter 1 showed how birders enter a flow state when on a nature walk, paying close attention to any sounds or movements in the woods, in their attempts to spot a bird.[2] Here, I add to the psychologist Mihaly Csikszentmihalyi's theory of flow, and I argue that birders experience a "residual flow" whenever they find and pay attention to birds, even if not on a walk. When they watch birds on their commute or out of their window while at work, birders can tap into the sense of calm and wonder that they experience on a bird walk.

Flow typically results from "a structured activity, or from an individual's ability to make flow occur, or both."[3] Csikszentmihalyi's example of a spontaneous flow experience is a good conversation at a dinner party. That type of spontaneous experience is not something people can train for, or something that people's "skill"—in this case, at conversation—can necessarily predict. Birding creates a unique opportunity for individuals to be able to tap back into their flow experience, in what I call "residual flow." Birders experience a true flow state when out birding, but they can revisit that state any time they see a bird. Birders' residual flow relies on their birding skills—their ability to pay attention to, spot, and identify birds. Even if for only a moment, birders experience the same kind of wonder while sitting in a train on their commute or while looking out the window at work as they do when out on a walk.

The idea that birders are "always birding" can refer to the time of year and how each season provides opportunities for different types of birding. On a walk in late July, Bill, the walk leader, kept reminding participants about the upcoming walks.

"This is the last walk until September, there's no walks in August," Bill said.

"Why aren't there any walks in August?" someone asked. "Is it just because everyone's on vacation?"

"There's just not much birding to be had in August," Bill explained.

"Fall birding is grand," one woman said.

"Winter is, too," Bill added.

"But spring is best because of breeding!" someone else exclaimed. Everyone then agreed that summer was simply not the best time of the year for birding.

Most often, when birders say they are "always birding," they mean that they are always on the lookout for birds. Andy talked about birding at the Audubon Convention hotel: "There's never a time when I'm not birding. I was looking at sparrows on the roof from our hotel window. Wherever they are, you're watching them." Other birders said that they enjoyed being the passenger in a car so they could watch out for hawks on the side of the road. Miranda said that she sees the same birds on her commute to work: "We have feeders at home, we have feeders at work, so there's kind of always that constant in the background. I actually posted on our Facebook page recently, about having a commuter bird, because I see Bald Eagles in the same spot every day on my commute. And so kind of whenever I'm out and about, I'm looking for birds. Kind of always in the background."

A flow state represents the peak experience of going on a bird walk—birders are so wrapped up in searching for any sign of a bird that they lose track of time and other senses. Here, the epitome of being in a state of "always birding" meant that birders experienced a residual effect of that flow state. Birders can experience the same calming, centering, and exhilarating feelings of birding anytime, simply by paying attention to birds wherever they happen to be. Cynthia said she birds as she walks to and from work or when she walks her dog, although she would not officially call this "birding": "I might not necessarily be birding, that might not be my purpose for being there, I'm walking to work, or I'm walking my dog, but the birds are there and now of course that I work for Audubon, I'm like, 'What do I hear? What do I see?' so I don't have my binoculars on me, I wouldn't call it birding, but they're there, and I'm seeing them and hearing them and appreciating them." It seems that having one's binoculars demarcates "birding" from "always birding," or paying attention to birds while outdoors or while looking out the window. Cynthia said she did not have her binoculars with her, but she was always paying attention to the birds she sees and hears as she walks to and from work.

Similarly, Tom, a retired carpenter who takes nature walks every day, said he is always looking for birds, even if he does not have his binoculars (or "glasses") with him:

> It gets to the point where you're in a permanent state of looking for birds. It's not like, "I'm going out birding today." Once you walk out the door, or even if you're inside, you're looking out the window. You're always, always looking for birds. You're never outdoors, even if you're walking down the path with, say, your spouse or your boyfriend or girlfriend and you're holding hands, you're still looking around for birds. You're outside, it becomes a way of life. You're never outdoors and not looking for birds. It's not to say, like, "I'm going to go get my glasses, and I'm going to go out, and I'm going to go look for birds, now." Once I walk out the door, I'm in a state of looking for birds. It does become a way of life.

To Tom, birding becomes a way of life, and as he said, birders are always looking for birds. When Tom and I finished our interview, left the diner, and walked outside, we both immediately scanned the trees and saw a Red-tailed Hawk glide into one. We both made sure the other had seen it, and I told him, "You were right—we are always birding!" Vivian said that she is "never not birding," even when at work: "I don't think I'm ever not looking for birds. Once you're a birder, you see them everywhere. You're never not birding. Even at work, something flies by the window, and you're like, 'Oh, maybe it's a Kestrel! Maybe it's a Peregrine!' It's usually a pigeon, but that's okay." Here, we get a sense of a key difference between birding and other naturalist activities: a birder doesn't have to take a trip to a zoo or a whale-watching tour far out into the ocean to see wild animals.[4] Birds come to us, whether we are inside a car or inside a building at work, even in the middle of a city.

Allen, who had seamlessly pointed out a Great Blue Heron in our interview outside at the National Audubon Convention, also described the Zen-like state he can enjoy when he taps into a state of "residual flow" and pays attention to the birds in the background of whatever else he is doing: "Like now, just sitting here, you can bird while we're talking because you can hear these birds out there—well, if the cicada wasn't making this noise. But I find I'm always sitting there talking to people, and in the background, I'm listening to Chimney Swifts fly overhead, or the Robins, or whatever. Some of the birds you can do [*looks away*]. I got distracted—something flew by." Allen demonstrated how birding "in the background" can become birding in the "foreground" when it supersedes what one is already doing. On a few occasions during our interview outside, he pointed out different birds that were flying or singing. I conducted several of my interviews outside, and birders were continually distracted by any birds that flew by or made any type of noise. Sometimes birders even got distracted by squirrels, much like what happens on a regular bird walk.

Barbara, the administrative director of a state-level Audubon organization, said she points out birds to her friends when they go hiking together, and her friends jokingly accuse her of always birding, even if she didn't bring her binoculars and is thus not "officially" birding: "I do just like to be out in the woods hiking. I don't have to be birding. I don't always take my binoculars when I'm out. But I'm always tuned in to the songs, and I'm listening. And when I'm with anybody else, I'll say, 'Ooh, there's a, whatever,' and they'll say, 'You're always birding!' And I'll say, 'No, I'm not!' [*laughs*] 'No, I'm not!'" Barbara laughed because she is, in fact, always birding, and her friends caught her at it. Like someone caught daydreaming, birders get caught birding when they should be completing other activities.

This residual flow might be unique to birders. Other hobbies, such as scuba diving or skiing, can create a flow experience for practitioners, but they require specialized equipment and locations to be able to enjoy that experience. Musicians don't experience the same feeling of making music when listening to music, nor do dancers when they watch a video of someone else dancing. Belting out a half measure or dancing a quick jig is not the same as an immersive flow experience. Csikszentmihalyi likewise argues that surgeons can experience flow when immersed in their work, but I certainly hope that surgeons don't spontaneously start cutting into people in order to enjoy a flow experience outside of work. Because birds are everywhere, and birders are always birding, birders can enjoy a residual flow experience at any moment. Even while at work, when they spot and pay attention to the birds around them, birders can briefly be transported back to the same feelings they get in the woods on the weekend, fully enjoying themselves. This ability to always be birding means that birders are always using their naturalist gaze to pay attention to the natural world around them.

BIRDS' MOBILITY AND OMNIPRESENCE

Birds' mobility facilitates birders' "always birding" and the naturalist gaze always being "on." Birds fly, and they bring nature, wilderness, and wildness with them wherever they go. Birds do this in ways other wild animals that people enjoy watching cannot—a person who enjoys whale watching will likely never get distracted by a whale flying past the window at work in a city. Birds also move about and bring nature with them in ways that other elements of the natural world cannot, such as unmoving flora.

On bird walks, in addition to showing participants how to find and identify birds, walk leaders also point out the various trees, shrubs, and bushes they encounter, especially if birds feed on those plants. Some groups even hold walks specifically for tree and wildflower identification. Thus, in my interviews I asked birders to explain to me what differentiates birding from other naturalist activities, like tree or wildflower identification. Catherine and Lenny, both avid naturalists who are also interested in dragonflies, explained the difference in a way reminiscent of how Theodore Geisel listed all of the places that a person could eat green eggs and ham:

ELIZABETH CHERRY: What makes birding different than other naturalist activities, like identifying wildflowers, or dragonflies?

CATHERINE: You can do it anywhere, you can do it in all seasons.

LENNY: To any extent, in terms of equipment, or how deep you want to dive into it.

CATHERINE: You can do it relatively inexpensively, or you can spend a lot of money. You can borrow binoculars.

LENNY: You can go by yourself, you can do it with a group.

CATHERINE: You can do it by yourself, you can do it with a group, there's resources at the library, there's resources online. There's apps for your phone, so you can learn the calls. Well, some of us can. I still have trouble with the bird calls. But Lenny does a good job of birding by ear. That was an interest that he took, so he would listen to the tapes.

EC: These are things that birding has that dragonflies don't?

CATHERINE: Dragonflies are very seasonal.

LENNY: I can't go out and get a dragonfly at the moment.

CATHERINE: At least not in New York. Without traveling somewhere.

To Catherine and Lenny, birds are available in all seasons of the year, birds can be observed with or without binoculars, alone or with a group, and there exist far more resources for birding than for dragonflies. In comparison with observing dragonflies, birding made for a much more accessible activity—anyone could do it, any time, with any amount of equipment. Dragonflies can be observed in summer, but not at other times of the year. Birds are everywhere, all of the time. In our interview, Rhonda jokingly compared birding to seeing one of the animals people see most often when out for a walk—dogs: "When you walk around, doing your daily activities, most of the animals you see will be humans. Maybe you'll see a dog. Maybe you'll see a cat, or a mammal. Maybe, depending on where you are, you'll see a horse. But birds are the most ubiquitous. You see them everywhere, all the time. And there's so many different species of them. They're all over the place. If I were to go 'dogging,' for example, not that many dogs, you know?" "Dogging," of course, does not exist. Rhonda's example proves another important point—even if you see several dogs while out on a walk, they will not represent different species. With birding, a birder can see several different species in any one day, in contrast to other naturalist activities.

When he goes "herping"—a reference to herpetology, meaning searching for reptiles or amphibians—Ethan said he often sees nothing at the end of the day. In contrast, birding means he will definitely see something: "When you go out herping, or you go looking for specific animals, you don't often see them. Often times you come out empty-handed. But birding, what bird watching does is, since birds are very abundant, and they're everywhere, and there are so many different species, and so many different types and varieties and behaviors, it's impossible to get, it's one of the reasons why you can't really get bored by going outside. And so birds, when I'm outside and there's no frogs or no salamanders or no snakes, I can always count on seeing, say, a White-eyed Vireo, or a Palm Warbler, or something high up in the trees." Ethan said he enjoys birding because he's guaranteed to see something, unlike herping or other naturalist activities. More than that, Ethan went on to say that seeing so many varieties of birds helps him understand that "nature has many different varieties" and that nature is "a very balancing, and fragile, ecosystem." The variety and ubiquity of birds allow such contemplation to happen more regularly, and allow the naturalist gaze to always be on, in ways that other naturalist activities do not.

Birds' mobility also brings wilderness and wildness with them, wherever they fly. Wilderness describes a place, and wildness represents a property of living or nonliving beings.[5] What does wildness mean to birders? Birders love birds' wild nature—birds are free. Birds defy our domestication, as Kay, an artist and art teacher, explained in our interview: "Untamed. Undomesticated. Even though there are some that are domesticated. Wild nature. What do they call that? The sacred and profane, they call that in art. So it's kind of like that thing. Where we've domesticated some things in the world, but that's not always a happy thought. And to think that there's this whole species that is going to defy that, and has that freedom."

The naturalist gaze focuses on wild animals in their natural habitats. Even though people have domesticated some birds as farmed animals or keep other birds as companion animals, birders appreciate wild birds' freedom as an integral element of their wildness. To birders, birds are free because they can fly. Birds' wildness means a lack of domestication. Birds are free and out of our control, as Ellie told me: "They're wild, and they're completely autonomous. They're just leading their lives freely. You kind of feel like you're looking at something out of your control, and that's, maybe that's one of the things that makes it so fascinating to do. That you're getting a glimpse of something that has no, you can't affect at all. Well, you can. And we know that we do, when we bird. But we think we're not affecting it. We think we're glimpsing something that's freely doing its thing." Ellie notes that birders fully understand that their presence affects wild birds' behaviors. This knowledge, that people affect birds and their shared ecosystem, makes up a key element of the naturalist gaze and is precisely why birders wear clothing that blends in to their surroundings and walk quietly on bird walks, so as to minimize their disturbance of birds. Ellie was initially speaking to birders' immediate presence affecting birds' behavior, but she also spoke to humans' effects on the natural world. "We think we're not affecting it," she said, even though she admitted she knew that all humans, including birders, affect birds.

Birds still act like free, wild animals, even when they're in the middle of humans' built environments. Birds' wildness transcends these environments, as they maintain their wild behaviors no matter their location. In our interview, Vivian further explained humans' encroachment on the natural world:

> With a lot of nature, I feel like it's their world, and we took over a whole hunk of it. But they're in that natural state, and we're dominating around it. They're still going about their business and trying to do their natural thing, so I think sometimes we're very imposing. And that goes to that whole perspective about them, when you're out in it, you get a different sense of self and place. Maybe in those moments, when you're walking through a field or whatever, you feel a little more a part of it, as opposed to someone who just tried to compartmentalize it and built things on it and changed it. I think they're still doing things pretty much the way they always did—we're the ones who keep changing it.

Like Ellie, Vivian also notes that birds and other wild animals are "trying to do their natural thing," even though they do so in a world that has been utterly changed by humans. Researchers in California, for example, have found that birds are now nesting earlier, as a way to avoid their chicks overheating in hotter temperatures due to global warming.[6]

Humans have changed animals' habitats for the worse, but animals can change humans' lives for the better. Vivian acknowledges this when she says people can "get a different sense of self and place" when walking in a field, looking for wild birds. Birders know that whenever they see birds, they are doing so in a spot that has been changed by human activity. Some areas, like national parks, attempt to preserve tracts of land from such disturbance. A sign in the Everglades National Park in Florida demonstrates how the Wilderness Act forbids certain types of human intervention in the parks. The sign reads, "Wilderness once encircled humans. Now we encircle it. The Wilderness Act intends that some federal public lands will keep their wild character forever. Roads, buildings, logging, drilling, most commercial activities, and mechanized transport are off limits here. Wilderness gives us a glimpse of what America once was."[7] Like the quotes from the birders earlier in the chapter, the signage in Everglades National Park also demonstrates that birds and other wild animals go about their business while humans change the world around them.

The "Wild Places" sign conflates wilderness and wildness by commingling wild animals with other elements of nature. In her work on the rhinoceros, the historian Kelly Enright clarifies that wildness refers to behavior, whereas wilderness describes space. However, they remain linked: "Though they [rhinos] can be removed, exhibited, and displayed outside the space of wilderness, their perceived wildness rests in their association with this landscape."[8] Enright was discussing the rhinoceros's exhibit as a part of special menageries throughout the United States and Europe during the nineteenth century. The rhinoceros, a land-based mammal, is found only in the wild in southern Africa and south Asia, and therefore its transport and display in other countries, such as the United States, represented the nature found in those distant lands. People from another continent would never see a rhinoceros in the wild, unless they took a special trip to southern Africa or south Asia. Birds, in contrast, regularly transport themselves from continent to continent via their migration pathways. Birds' wildness relies less on representing different lands and more on birds' ability to move from landscape to

FIGURE 5. "Wild Places" sign in Everglades National Park (photograph by author).

landscape. Even when human activities disturb birds' habitats, birds can still retain their wildness as long as they can fly.

Even though development has overtaken most wilderness areas, birds still maintain their wildness by adapting and persisting in people's built environments.[9] Birds occupy a liminal space—they bridge the gap between wildness and civilization, between nature and culture. Enright says that "wildness is something 'out there,' something removed from civilization," but birds maintain their wildness even in the midst of cities.[10] Cities do not provide natural habitats for birds. Concrete buildings may resemble the rocky cliffs that are the natural habitat for Rock Pigeons, but pigeons survive, and have learned to thrive, in cities—they do not originate in cities. Birds reclaim their wildness by persisting in human-dominated spaces, as Chloe explained in our interview: "Just how extremely adaptable they are. It's like the House Wren will, they'll build a nest in a cowboy boot that's been left out on the porch for too long. Or they'll, House Sparrows, the way they'll build a nest in a Wal-Mart signage, a neon sign. Swallows who will build their nest underneath a major freeway. I mean, they just, they persist even though we're there, with all our stuff. I find it to be encouraging."

Birds express their wildness in the middle of human-dominated spaces, even though such spaces are not "wilderness," as Snyder reminds us: "Wilderness is a place where the wild potential is fully expressed, a diversity of living and nonliving beings flourishing according to their own sorts of order."[11] Even if birds cannot "fully express" their wild potential, birds'

FIGURE 6. European Starlings on a power line (photograph courtesy of Dave Saunders).

wildness can still be found in cities and other human-dominated places. Even though birds live in humans' built environments and nest in neon signs, they still retain some of their wildness simply by virtue of being a bird and by conducting their bird activities, such as breeding and nesting, in the midst of human development.

William Cronon wrote, "Wildness (as opposed to wilderness) can be found anywhere."[12] This distinction between wilderness as a property of a place and wildness as an element of a living being is crucial for understanding the difference between wild plants and wild animals. Would we understand a wildflower that grows in a crack in a sidewalk in the same way as birds that nest in a neon sign? I would say yes, and I imagine many of the authors mentioned in this chapter might say the same. But as I demonstrated in chapter 2 with Walter Benjamin's natural aura, there still remains a key difference between wild plants and wild animals, and especially wild birds, which can fly away and live freely on their own. Birds' ability to fly helps them reclaim their freedom, even after they have been in captivity, as Kay told me in our interview: "You know about those parrots that were let free and are in Brooklyn and Queens. And look at those guys, they're kind of walking the space between those two worlds of we were domesticated, and we're still hanging around people, so we're not trying to escape, but we're reclaiming our wildness."

Birders' naturalist gaze encourages them to view birds and people as cohabiting a shared ecosystem, and thus birders avoid the problems that come with viewing oneself as entirely separate from nature, as Cronon warns: "If the core problem of wilderness is that it distances us too much from the very things it teaches us to value, then the question we must ask is what it can tell us about *home,* the place where we actually live. How can we take the positive values we associate with wilderness and bring them closer to home?"[13] Birders precisely do bring the "positive values we associate with wilderness" to their homes when they engage in backyard birding and feeding of wild birds, which I discuss in the next section. However, Cronon's writing cannot help us fully understand this relationship between birders and birds, as he tends to conflate wild places and wild animals: "When we visit a wilderness area, we find ourselves surrounded by plants and animals and physical landscapes whose otherness compels our attention. . . . The same is less true in the gardens we plant and tend ourselves: there it is far easier to forget the otherness of the tree. . . . The special power of the tree in the wilderness . . . can teach us to recognize the wildness we did not see in the tree we planted in our own back yard."[14] In contrast to a tree or a landscape, birds are mobile—the most mobile of wild animals. Wild animals, and particularly birds, present a wholly different phenomenon than trees and landscapes. Cronon lumps them all together under "wilderness" or "wildness," but it is crucial to separate living animals from plants and rocks, not only for the purpose of avoiding anthropocentrism and speciesism but also for the purpose of better understanding how birds, in their mobility, can bring a bit of the wilderness to us.[15] Birds bring wilderness to us whenever and wherever we see them—nesting in neon signs, in a city, outside our window at work. We do not apprehend birds only while on nature walks in forests. Therefore, we need a way of thinking about birds that is qualitatively different from how we think about trees, because birds are qualitatively different beings.

Birders think about birds differently than how they think about other elements of nature. Birders expand their understanding of nature, since they know that birds are found in places other than pristine wilderness areas. Birders even find birds in grocery stores, as Joy noted in our interview: "It's a lot easier to be a birder than it is to be focused on another species, because birds are everywhere. The flowers don't move, the turtles don't move too much. You can find birds everywhere. You can find them in the parking lot. There are even some stores, if you go in a big box store, chances are there's at least one bird in the store, even if it's a grocery store." Earlier, Ethan said he enjoyed birding while out on his herping walks, because birds were more observable than reptiles or amphibians. Joy took this sentiment further to note that birds really can be found everywhere—not just in every forest or on every nature walk, but also in less "natural" places like grocery stores and parking lots.

This ability to find and appreciate birds everywhere allows birders to "imagine a civilization that wildness can endure."[16] When birders notice birds, they notice the wildness that such birds bring with them wherever they go. Just as the rhinoceros brought images of far-off wilderness and dangerous wildness with him wherever he was taken by his captors, birds bring a sense of wildness with them, wherever they fly.[17] The captive rhinoceros brought an illusion of wilderness with him; the environmental philosopher Thomas H. Birch would call it a simulacrum of wilderness.[18] To Birch, governments create wilderness areas in order to more beneficently dominate the wild beings in those lands. Similarly, the display of a wild animal as large and ferocious as a rhinoceros is equally a display of people's conquest over animals, and the wildness that such a captive animal exudes is a mere simulacrum of the real wildness that it could have in a world without human intervention.

Birds are more observable than other wild animals because they are everywhere, even in grocery stores and parking lots. Since birds are everywhere, this means people can bird everywhere. Of course, birders do not hold bird walks in grocery stores. Their naturalist gaze is simply always on, allowing them to appreciate birds anywhere and everywhere. To these birders, birds' ubiquity means they can observe birds anytime, anyplace, anywhere, as Cynthia described her experiences birding in a park near her home and work in a major city: "It's enlightening, even for people who are very experienced birders, to come birding in a city where the birds tend to not be as shy, and since it's not a dense forest, but more like this scene where you have sort of a mowed grass and then canopy. If I'm standing here and I know there's a bird in this tree but I can't see it, well, I walk around to the other side of the tree. And that's not something they can do in a lot of the habitats where they're used to birding. It's a great place for beginners to learn their birds, but I think we also surprise a lot of long-time birders. . . . We get to see a lot more than I think that people realize." Birds are everywhere, even in the middle of cities. Like Cynthia, Sue said she has birded in cities, even outside of parks: "I've birded in New York City, not Central Park, but in the main drag." Andy said he traveled for one of his other hobbies, and whenever he has free time, he can always bird, because "you can do it anywhere. I'm into cars, too, and I go with my car clubs and travel around, and no matter what, there's downtime, I can bird. It's doable."

Wherever birders travel, even in a city, they can always find opportunities to go birding, or even just observe birds. People typically view wildness as outside or beyond the boundaries of human culture.[19] Birds bring their wildness into human culture, into human places, and into birders' consciousness. Birders try to share this consciousness with others. Birds' omnipresence also means that birders can teach others about birds. Several of the birders I interviewed worked as environmental educators, and they said the ubiquity of birds helps them teach about various science or environmental topics. As Kristina said: "Kids can see birds anywhere. That's an animal that you know you'll always see. So, it's easier to connect than maybe something else that, you know, a deer or a fox or a snake or a turtle. I love the idea of nature and inspiring kids with it. But using birds kind of as a portal because they get to see them so often." Kristina used birds to inspire children to learn more about nature, and the ubiquity of birds helped environmental educators teach children about nature and science concepts.[20]

Birders also sought to teach adults about birds and environmental issues. Dawn said that birds being everywhere meant she could talk about conservation anywhere and everywhere, even in a parking lot: "I always say, as an environmental educator, I love using birds as a portal into conservation because I can be anywhere. I can be in the middle of Dulles Airport parking lot and see birds, or in the middle of Manhattan and see birds, or in Yellowstone National Park and see birds, so you know, you can really engage people in nature and conservation through or with birds. So the fact that they're everywhere, they have unique behaviors, they do amazing things, they migrate, they have eggs, they're just really special creatures." Birds' unique behaviors and characteristics, combined with their ubiquity, always provide opportunities to talk to others about birds and environmental issues. Catherine said this gave birding organizations an advantage over other wildlife conservation organizations: "Birds are an opportunity, because they're visible. World Wildlife Fund uses a panda. Pandas are cute, but they're all in zoos, or they're in China. So, it's recognizable. But it's not something that people are going to see unless they visit a zoo, or get a World Wildlife Fund sticker." In comparison with Enright's rhinoceros, panda bears are much cuter, but they essentially serve the same function as the rhinoceros when representing wildlife. Both pandas and rhinoceroses represent distant wilderness areas and exotic wildlife species that most people will never see outside of captivity. Giant pandas are native to only a few mountain ranges in central China, though 49 of the 1,864 living pandas live in captivity outside of China. Despite their remoteness, giant pandas represent the World Wildlife Fund and they represent endangered animals (although in 2016, the International Union for the Conservation of Nature reclassified pandas from endangered to vulnerable). In contrast, wild birds are everywhere, able to freely move from place to place, and therefore birds provide a more realistic and accessible representation of wild animals in need of conservation. Since birds' ubiquity facilitates birders' naturalist gaze always being "on," and since birds' mobility brings wilderness and wildness wherever they go, birders can always tap into nature and share their appreciation of nature with others.

BACKYARD BIRDING

In his generative writings on wilderness, the environmental historian William Cronon warned that imagining wilderness as far away results in our ignoring the nature that can be found in our own backyards: "Idealizing a distant wilderness too often means not idealizing the environment in which we actually live, the landscape that for better or worse we call home."[21] Birders provide a perfect counterexample to Cronon's concerns. By appreciating common birds, as described in the previous chapter, birders imagine, or socially construct, nature to be everywhere. This chapter thus far has shown how birds' mobility and ubiquity help people see wild animals in the wild and connect to wilderness and wildness through birds. Birders are "always birding" and thus their naturalist gaze is always "on." Since birders are always watching birds, wherever they are, and since birds bring wilderness and wildness with them wherever they fly, these practices culminate in birders seeing, and creating, a bit of wilderness in their own backyards. Every birder I interviewed and observed watched birds in their own backyards, even those who lived in apartment buildings in cities. They also created habitats for birds by planting native plants and feeding birds in winter.

Birders experience the wonder of seeing birds anytime and anywhere because they always pay attention to the outdoors. Since birders are always looking for birds, they could spot uncommon birds outside their windows, even when they lived in the middle of a city. Cynthia, who lives in the city, said, "I don't really have a yard, per se," but she said she is luckier than most city dwellers since she had a back patio that allowed her some outdoor space: "My bathroom window in the back looks out onto my neighbor's patio, but there are a couple of trees coming out of the sidewalk randomly—I'm sure they're not on purpose—and I've seen a Scarlet Tanager back there, I've seen Eastern Towhees back there. You would be shocked! It's not by any means a forested area. It's like this tree accidentally grew out of the sidewalk, and it's there, and during migration I might just be brushing my teeth and I'm like, 'Holy cow! There's a Towhee!'" Cynthia had to pay attention to what was happening outside her window in order to enjoy and experience the excitement of seeing these birds. Since birders also appreciate even the most common of birds, the fact that tanagers and towhees were less common only added to the excitement of seeing a bird out the window. The key is that Cynthia and other birders pay close attention to what happens in their backyards, and thus, in stark contrast to Cronon's admonitions, they can appreciate a bit of wilderness and a bit of wildness in their own literal backyards.

Amy described her development as a birder and how she learned to start paying attention to what was happening in her own backyard: "I feed birds in my yard, and it's been a very nice carryover between being able to identify the birds at the Point, and just being more in tune to things in my yard. I'm noticing things coming into my yard now that must have always been there, I just never noticed them. I have migrating warblers coming into my yard, like Common Yellowthroats. I never even noticed that before. But now I'm tuning into them. . . . It's what we're talking about, that people sometimes don't appreciate what's in their own backyard, and how you can bring things in." Amy said she is "much more conscious of what's in my yard now," and she planted three native shrubs along her driveway, which bring Cedar Waxwings to her yard. They come to her house, she said, "not for any other reasons. Because I have the fruit that they want." Creating this backyard habitat for birds allows her to "sit in your own back yard and again, to be able to just tune into the incredible diversity of bird life that I have." Birders experience the excitement and wonder of seeing both common and uncommon birds in their own yards, and they can create a welcoming habitat to create a bit of wilderness and wildness in their own yards, which the birds bring to them.

Birders enjoy birds in their own backyards whether they feed birds or not. They can always enjoy the birds that they see and hear in their immediate area. However, birders don't necessarily go out to their backyards to list and identify the birds that they see—they can simply enjoy the fact that birds are around them, as Andy explained, when contrasting his backyard birding to the bird walks he used to lead for Audubon:

> I bird my own property, so that's about the extent of my birding. I don't go out and organize walks that much anymore. I used to lead them for a local Audubon chapter and things like that. For me, it's just welcoming the morning and enjoying the beginning of the day, and especially during the warbler migration season, just seeing what's out there, and listening to sounds. I usually spend no more than an hour outside in the morning, the beginning of the morning. And that's birding for me, now. It used to be pretty hardcore where I would lead weekend walks, lead groups of fifteen people. And help them identify birds. But that's, my birding today is very sort of home-based, casual, on my own time kind of thing. Very enjoyable. It's relaxing.

Several of the birders I interviewed led birding walks, like Andy, and they differentiated between the formal, organized birding walks and the informal birding they did in their own backyards, in a way similar to their differentiation between "official" birding walks where they bring their binoculars and the informal "always birding" that they do when they tune in to the nature around them. Backyard birding represents another type of informal birding that birders can enjoy since their naturalist gaze is always "on."

Even though backyard birding represents a less formal type of birding, many birders keep lists of the birds they see in their yards, as Mona said, "Of course I keep a life list for my yard." Whether they keep lists or not, these birders all enjoyed the simple pleasure of watching birds in their backyards, as Dawn said, "Because I do a lot for work that has to do with birds, for just kind of pleasure, it often is much more infrequent and informal. So sometimes it's literally watching birds in the backyard. I live in rural Pennsylvania, so we have a pretty good variety of species in our backyard." Although backyard birding can be quite informal, the birders I interviewed said they still paid attention to even the most common of birds in their backyards, as Toby did with the robins and cardinals that nest in his backyard: "I have bird feeders, and I have a lot of birds in my house, and I really enjoy, I sit on my deck, and I enjoy watching the birds at my house. When I'm out in the field, I focus on other birds, because I'm out in the field. So it may seem like I'm not paying attention to that robin, but I really do pay attention to the robins that are nesting in my yard. The cardinals, and all that, like they nest, I have nests all over my yard."

Whether passively enjoying the birds by sight and sound in their yards or actively keeping lists of the species they see in their yards, these birders all appreciated a little slice of "wilderness" that the "wildness" of wild birds brings to them, in their own yards. Since birds are wild animals, and birds are mobile, birds brought their wildness to birders' urban, suburban, and rural backyards. In these ways, birders exemplify the more wide-ranging conceptions of wilderness and wildness that Cronon and Snyder encourage, rather than seeing wilderness only as formal, protected public lands: "Wildness is not limited to the 2 percent formal wilderness areas. Shifting scales, it is everywhere: ineradicable populations of fungi, moss, mold, yeasts, and such that surround and inhabit us. Deer mice on the back porch, deer bounding across the freeway, pigeons in the park, spiders in the corners."[22]

Birders know their backyards do not represent true wilderness areas, but they nevertheless attempt to create a healthy habitat for birds in their backyards. As Amy said earlier, it's no mistake that birders have wild birds in their yards—it's because birders have intentionally created bird-friendly habitats for the birds. Every birder I interviewed who had a yard created a bird-friendly habitat in their yard. Such habitats included a variety of elements, as Dawn explained: "We plant a lot of native plants, and we have bird feeders up in the wintertime, and hummingbird feeders up this time of year, and nest boxes for Bluebirds and Tree Swallows, and we definitely try to create the habitat, so we've reduced quite a bit of our mowed lawn and created habitat." The birders I interviewed had a variety of bird feeders, bird baths, native plants, nest boxes, and the like all throughout their yards. On walks, people often shared their favorite methods for attracting special birds, like orioles: some people swore by oranges cut in half, others bought special oriole feeders, and one person said she used grape jelly to attract orioles.

Learning to create bird-friendly habitats was so important to birders that some of them underwent special training for it: Mona became a "habitat steward trainer," meaning she trains others to create backyard habitats. I describe birders' environmental emphasis on native plants in more detail in chapter 7, but here, the focus is on how creating healthy backyard habitats allows birders to enjoy more birds in their yards while also helping birds. Miranda runs a sanctuary associated with her Audubon organization, and she remarked that people always wondered how they attracted so many birds to the area. Miranda explains that it's easy; just provide good habitat:

> Birders tend to have this aha moment, where they see a bird, and they identify it, and then all of a sudden, they're aware. It's like someone's taken off earmuffs and sunglasses and they see them everywhere. People come to our wildlife sanctuary, and they're like, "Why are all these birds here? What do you do to make them come here?" And we're like, "We don't do anything special that you couldn't do at home. We have bird feeders, we have good habitat. We're removing invasive species. We're managing to improve the bird life, and anyone can do that at home." But it's like, they come, and they're like, "Whoa, so many birds!" They've been here. They've always been here. I think it's just, how do you spark that aha moment? How do you cultivate that? And grow it?

Miranda said her organization holds the birdseed sale fund-raiser "because the more that you appreciate the birds in your backyard, the more that your eyes get opened up to the birds that are more rare. Or harder to find. Or harder to appreciate. Or maybe you do have to take a special trip to see. Or maybe you have to take a minute and really listen, to hear, to pull out that species." Miranda's comments also demonstrate the bigger picture of why providing good habitats matters—appreciating birds in your own backyard, she reasons, will provide a starting point for people to notice more birds, appreciate more birds, and help more birds, thus also helping the environment that we all share. In these ways, birders further develop their naturalist gaze, which helps people view themselves and wild animals as part of a shared ecosystem.

This chapter showed how birds' mobility and ubiquity help people view wild animals in the wild and connect to nature. Birds help people see nature as anywhere and everywhere, rather than as separate from human society, or remote and untouchable. Birders are "always birding," always tuning in to nature, wherever they see birds. Birds bring their wildness to a variety of places, including parking lots, airports, big-box stores, city streets, and outside people's windows at work. In doing so, birds help birders conceive of "wilderness" in a broader sense, incorporating their own literal backyards into their expanded view of nature. By always birding, by seeing birds everywhere, and by inviting birds into their own backyards, birders develop their naturalist gaze to view themselves as sharing the natural world with birds and with other flora and fauna. In the next chapter, I show how birders also use their naturalist gaze to evaluate this natural world.

5 ❧ GOOD BIRDS, BAD BIRDS, AND ANIMAL AGENCY

CONSIDER THE TWO following scenarios, both taken from different bird walks I attended with local birding groups. On a bitterly cold February morning, a group of birders marches through the woods, on the search for a Barred Owl they heard was in the area. We find the owl, but only one man gets a good look before the owl flies away. Just when we start debating whether to go farther into the forest to try to find the owl, one of the birders points out a Red-bellied Woodpecker in the tree next to us. Everyone trains their binoculars on the bird, and then Tom exclaims, "Hold on, that's a Yellow-bellied Sapsucker! That's a really good bird!" We spend the next few minutes looking at field guides to understand the differences between Yellow-bellied Sapsuckers and Red-bellied Woodpeckers, our quest for the Barred Owl put on hold to appreciate this "really good bird."

On a much more pleasant morning in late June, as another group of birders waits to begin their walk, the participants pepper the walk leader with questions: "Is it Starlings or Grackles that are tearing up all of the houses in the Catskills?" one woman asks. "I don't know if it's Starlings or Grackles, but let's blame Starlings because they're gross," Arnie, the walk leader, replies.

What makes the Yellow-bellied Sapsucker "a really good bird," and what makes Starlings "gross"? Why are some birds "good" and other birds "bad"? Throughout my fieldwork, I kept hearing the phrase "that's a good bird," and I kept observing the quick and often harsh judgments of other birds deemed "bad," so in my interviews I asked birders to tell me about their favorite and least favorite birds. My observations, and my interviewees' responses, help us better understand these evaluations and animal agency.

The naturalist gaze is evaluative: Birders categorize the habits of particular species of birds on the basis of their knowledge of the species' origins and their role in the ecosystem. The naturalist gaze is integrative: birders also take into account humans' intrusions into birds' natural habitats. "Good" birds are typically beautiful or uncommon, but more importantly, they belong in the ecosystem where birders observe them. Other birds are "good" because they provide ecosystem benefits. "Bad" birds exercise agency in ways that detriment the local ecosystem, including harming other birds. Returning to those "gross" Starlings, Arnie explained his judgment to the group: "Starlings are very invasive, and they'll kill other birds. They'll follow woodpeckers and use the nest holes that woodpeckers make, and then kill the woodpeckers. People are making Bluebird boxes now, to help save the Bluebirds, and Starlings will kill the Bluebirds and take their nests. So, I have no problem blaming Starlings."

In this chapter, I first present sociological understandings of animal agency. Then, I look at what makes "good" birds good: how they fit into an ecosystem, and how they exercise their animal agency to provide ecosystem benefits. I extend this theoretical discussion using examples of three "bad" bird species: European Starlings, House Sparrows, and Brown-headed Cowbirds. I show that the distinction between "good" and "bad" is not so clear-cut. The "bad," invasive, nonnative House Sparrows and European Starlings avoid full blame, because birders know that the primary criteria that make them bad—their invasiveness and nonnativeness—are not their fault. People are to blame for bringing starlings and sparrows to North America in the first place. In contrast, birders view Brown-headed Cowbirds, a native species, as "bad" because of their parasitic behavior. Cowbirds are brood parasites that lay their eggs in smaller songbirds' nests, thus endangering those songbirds through their behavior. The naturalist gaze blames people for the habitat destruction wrought by certain invasive species, and it implicates birds' animal agency when native bird species endanger other birds. These birds, and birders' evaluations of them, help expand our notions of animal agency and its intersections with animal instinct and human agency.

ANIMAL AGENCY

Birders are not the only people to evaluate animals. The sociologists Arnold Arluke and Clinton Sanders developed the "sociozoologic scale" to describe how people rank animals on a scale of how well they fit into human-dominated societies.[1] At the top of the sociozoologic scale are humans. Down the scale from humans are our compliant pets, and below them are lab animals, who serve as mere "tools" for scientists. According to this scale, companion animals and lab animals are "good" animals, since they know their place and willingly stay in it.[2] Good animals reify hierarchies in human society, and they are rewarded for it. Good animals, such as pets, are also quite visible in human society because they participate in it on human terms. Further down the sociozoologic scale, "bad" animals hold a low status since they don't submit to the social order. "Bad" animals intervene in human society and don't accept their subordinate place.[3]

The sociozoologic scale provides a useful starting point for understanding birders' evaluations of "good" and "bad" birds, but it does not fully incorporate animal agency into its ranking system, nor does it consider systems beyond human societies, such as ecosystems. As I demonstrate in this chapter, birds' goodness or badness primarily relates to how their actions fit into, benefit, or detriment the local ecosystem.

In sociology, agency refers to free will or making choices, especially moral or ethical choices. When sociologists turn their attention to animals, animal agency generally refers to the ability to cause change, and it does not necessarily imply self-consciousness, morality, or other aspects we attribute to human agency.[4] We can understand animal agency as the sociologist and foundational animal studies scholar Leslie Irvine does, as "the capacity for self-willed action."[5] Similarly, the sociologists Lisa Jean Moore and Mary Kosut define bee agency as "the bees' free will and ability to decide on a course of action."[6] Birds exhibit these forms of animal agency all the time, through their actions, especially through their ability to fly.[7]

Birds offer an opportunity to further our understanding of animal agency. Birds live in a world wholly affected by humans, but birds remain free because of their ability to fly. Most of our conceptions of animal agency come from studies of animals who are controlled in some way by humans—companion animals, lab animals, farmed animals, or animals kept in captivity for entertainment in zoos.[8] In contrast, we know little about the agentic interactions between humans and wild animals, such as birds. When scholars study human-wildlife interactions, they tend to focus on the danger posed by wild animals to humans, and they tend to ignore times that such "wild" animals don't harm humans.[9]

The naturalist gaze takes all of these elements into account. Birders understand birds as free wild animals, but at the same time, birders know their actions affect birds. Through the naturalist gaze, birders grasp the interconnectedness of birds and humans in a shared ecosystem, in which humans affect birds, while also seeing how birds themselves affect the ecosystem. Birders thus view birds through a combination of human agency and animal agency as self-directed action, all taking place in a shared ecosystem.

As discussed in the previous chapter, birders also appreciate birds' agency, their ability to fly, as a key element of birds' freedom and wildness. Birds' ability to fly, to choose to fly away when they could stay in one place, is a simple starting point for understanding birds' animal agency. Cynthia told me that birds' ability to fly away made birding more challenging than other naturalist activities: "I would say it's more challenging. Just because a tree stands still, and I can walk up to it. I can touch it, and look at it, and smell it, if I'm trying to identify it. Birds are a little trickier. . . . Whereas, you know, a plant or a flower, you just walk up to it." Every birder I interviewed noted this primary difference between birding and other naturalist activities—birds can fly away, whereas trees, rocks, and wildflowers cannot. In this sense, birding resembled hunting, in that there was a thrill in observing birds that could fly away at any moment.[10] But unlike hunting, birding has only observation as the end goal, not killing or capturing the bird.

Birders' understanding of flight as an element of birds' agency helps birders contemplate animals as individuals. Wild birds invite a different way of understanding animal selfhood than do companion animals. Leslie Irvine contends that when people visit animal shelters and look at the adoptable cats and dogs, part of what they're doing is trying on different selves of their own.[11] Colin Jerolmack's pigeon flyers used pigeons to create their own selves, and they did not appear to contemplate the birds' selves.[12] Because wild birds can fly away, when birds deign to stay in one spot and return the gaze of the human, birders contemplated the birds' state of mind and self, as Chloe explained: "They [birds] don't stay still. Wildflowers are so polite that way. They don't run off or do things like that. They're just, they have so much more personality, you know? You can almost have to imagine yourself thinking, 'What are they thinking?' They look at you with that one-eyed look thing, and you have to think that they see you, and what are they thinking? I know it's a little different obviously than most other things. I don't ever have that thought with other things." By "other things," Chloe was referring to other flora and fauna that she observed in her other naturalist activities, like identifying wildflowers. As discussed in chapter 2, the gaze represents power relations. Those doing the gazing typically hold some form of material or symbolic power over those being gazed upon. When wild birds return the gaze of birders, they doubly subvert the naturalist gaze by not flying away, and by returning the look, showing the birders that they, too, have been noticed. Of course, this returned gaze brings more contemplation than fright, as Chloe said she wonders what the bird is thinking. Were it a lion, or even a Cassowary (the third-tallest and second-heaviest bird in the world, with a reputation for attacking humans), the birders would undoubtedly feel less contemplative and more fearful.

Birds also exhibit agency in their interactions with birders. Some birders who bird by ear can also imitate bird calls. On one walk, Diane, the leader, whistled at a Mockingbird, who whistled back. They went back and forth a few times, and then Diane said, "Sometimes you can really piss them off!" Other birders talked about the need to avoid getting too close to owls' nests, so that they would not be attacked by the birds protecting their babies. While poison ivy can harm naturalists, it does not do so with the same sense of agency that birds and other wildlife can employ.

Even in wild systems, humans can affect animal agency by capturing wild birds and taking them to faraway lands, into ecosystems where those birds do not belong. This practice was common among early naturalists, before people understood the concept of ecosystems.[13] Now, several of those birds (and other species) have proliferated, becoming not just nonnative species but invasive species in North America. To be an "invasive" species (of animal or plant) means that the species proliferates so quickly that it becomes a danger to animals and/or humans. Invasive plant species crowd out and kill the native plants that native animals need for sustenance. Invasive animal species compete with native animal species for food and breeding grounds. Since invasive animal species have no natural predators in their new environment, their overpopulation also damages humans' built environments, such as nesting in buildings or taking over agricultural fields.

Using the naturalist gaze, birders evaluate invasive species of birds, taking into account the human agency involved in bringing them to North America. The evaluative element of the naturalist gaze judges the flora and fauna in an ecosystem on the basis of scientific information, and the concerned element of the naturalist gaze means it is interested in the health and well-being of wildlife in its natural habitat. This combination of human and animal agency, along with the scientific understanding of invasive species in an ecosystem, sets the stage for birders' complex evaluations of birds.

GOOD BIRDS

Many of the ways that humans use animals fundamentally serve to rank animals as "good" or "bad." The U.S. Department of Agriculture (USDA) grades the meat of farmed cows as choice, prime, or select. The Westminster Kennel Club Dog Show proclaims a Best in Breed and a Best in Show. People race horses in the Kentucky Derby, the Preakness Stakes, and the Belmont Stakes in an attempt to win the Triple Crown. In all of these cases, the winner is the human, who enjoys the fame and fortune of winning a contest, the money made from animal agriculture, or a meal made from another animal's flesh. The nonhuman animal ends up being the loser, forced to breed more prize-winning or meat-producing offspring for future use by humans. These uses of animals align with Arluke and Sanders's sociozoologic scale, in that people typically rank good animals according to their utility for humans—other animals serve humans as companion animals, or as "tools," as in the case of lab animals.[14] Wild birds do not fit in either of these categories. Wildlife observation, even when the animals are in captivity, does not involve such rankings. There is no "Best in Zoo" or "Best in Forest." And yet, birders assess wild birds, using the naturalist gaze. Birders deem birds "good" based on their behavior, including how they fit into an ecosystem and how they exercise agency to provide ecosystem benefits.

To Rhonda, the physical beauty of Atlantic Puffins initially drew her to appreciate the birds, as she said, "I like black and white, and I think their bills are just amazingly beautiful." Rhonda said Puffins' beauty appealed to her aesthetic sensibilities as a sewer and a quilter. What truly made Puffins "good," though, was how their beauty fit into a larger natural order: "That bird's bill is just a marvel of nature. Whoever invented that is just a genius. That's another reason I go out. It's just amazing what you see in nature. Whoever thought of that? And the fact that it all fits in, it all intersects like a puzzle, instead of the man-made things that we do, we make, we create, don't work so nicely and cohesively and interdependently together. Whereas nature does. It's pretty amazing." The naturalist gaze is integrative, viewing humans and birds as part of a shared ecosystem. It also integrates birds into the ecosystem, understanding why birds' different bill shapes have evolved for their different food sources, how their colors match their surroundings, and the like. Rhonda saw birds as part of a cohesive, interdependent element in the natural world, and their physical beauty formed part of that.

Diane talked about how the American Kestrel's colors drew her to the bird at first, when she was younger, and then she learned more about the bird, which fueled her fascination: "It's a falcon, it's incredibly colorful. There's just something about it. It's this small package of fierceness, has all these colors. The first time I saw one, I was maybe 10, and I thought it was a parrot, because it was orange and blue. And then to find out it was a bird of prey was pretty cool." To Diane, Kestrels were "good" birds first because of their physical beauty. Then, she learned they were birds of prey, and their behavior as a bird of prey then became a key element of her evaluation of them as "good" birds.

Birds of prey play a role in the ecosystem, even though celebrating predation is not the most popular topic of conversation among nonbirders. Birders appreciate even distasteful elements of bird behavior when such actions ultimately benefit the ecosystem in which the birds live. Using their naturalist gaze, birders see the benefits that certain birds bring to a particular ecosystem, even if their behavior may seem unpleasant. For example, Miranda discussed why even carrion eaters deserve our appreciation, since their behavior keeps us from wading through dead bodies: "They provide ecosystem benefits to us, like vultures and crows and ravens and even Red-tailed Hawks and Bald Eagles, they eat roadkill and other carrion. People will say, 'Oh,

I hate vultures,' and I'm like, 'Imagine a world without vultures.' I know it seems unseemly, what they do, but we would be up to our kneecaps in dead stuff if we didn't have them. They eat pests, and bugs that we don't like, and they pollinate things, and they spread the seed of all kind of plants."

The environmental philosopher Thom Van Dooren tells the story of three species of endangered vultures in India, and how "good" birds that provide a "service" to people by scavenging carrion can become endangered because of humans' environmental neglect.[15] India remains the last habitat for three species of vultures—the Oriental White-backed Vulture, the Long-billed Vulture, and the Slender-billed Vulture. Other areas in Southeast Asia used to house these birds, but after food shortages in these areas, the birds remained only in India, where they thrived. India proved to be an ideal place for these vultures to prosper, since it holds one of the largest populations of cattle in the world, the majority of which are not eaten by humans. When the cows die, people take them to carcass dumps or leave them at the edge of villages, to be eaten by the vultures. Additionally, the Parsee community in India uses vultures in its death rituals. The Parsee community lays out its dead in *dakhmas* (towers of silence) specifically to be consumed by vultures, believing that burying or cremating the dead would pollute nature.[16]

However, these vultures began dying off, owing to the vultures eating cattle (and some humans) that contained the anti-inflammatory drug diclofenac.[17] Farmers give old and sick cattle the drug to keep them working longer, and when they die, vultures eat the cattle and die from consuming the drug, which is toxic to vultures. As the vultures die out, fewer are left to eat the carcasses of the cattle, and many carcasses lay unscavenged in India, thus polluting the area. We need those "good" birds to perform these unsavory duties.

Truly free wild birds, such as vultures, show where animal agency and instinct meet. Carrion eaters have evolved to eat dead bodies, and thus it is instinct, not choice, that drives them to eat the dead. But in robust ecosystems, carrion eaters have more choices of food sources, and they can choose to scavenge the carcasses of livestock or wild animals.[18] In healthy ecosystems, carrion eaters can exercise agency and choose to perform this "service" for humans.

Birders appreciate even the most distasteful behaviors of free wild birds when they fit into an ecosystem. We rarely see such appreciation of wild animals' freedom and agency. Typically, the wild animals people celebrate have submitted to humans' control. The sociologist Colin Jerolmack studied various ways in which humans interact with pigeons, including feeding pigeons in city parks or keeping and racing pigeons in organized competitions.[19] Part of the draw of these sorts of interactions with pigeons was people's control over nature. City park feeders marveled at how they could entice the feral pigeons to perch on their hands, and pigeon flyers loved their pigeons precisely because they produced and controlled them. These pigeons moved up the sociozoologic scale because they were able to be controlled by people. Birders, in contrast, appreciate the freedom and agency of wild birds, and they marvel at birds' behavior in the absence of human intervention or control.

Birders also appreciate birds simply because they belong in a certain ecosystem, whether they provide ecosystem benefits or not. Migratory birds, which pass through an area only twice a year, do not "belong" permanently in the areas through which they migrate, but birders expect to see them in spring and fall migration. Seeing these "good" birds pass through means that the local ecosystem provides enough resources for them to stop over during their migration, which birders appreciate. Birders' excitement over seeing these migratory birds also highlights birders' appreciation of birds' agency in their freedom, mobility, and flight.

As noted earlier, when I first started joining bird walks I noticed that birders often say "That's a good bird." On walks, birders use this phrase to describe uncommon birds they are delighted to see and thrilled to add to their species list. These birds may be permanent residents and simply uncommon because they are difficult to see, but often these "good" birds were migrating through, and this was the only chance to see them that year. For example, as Luis prepared us for a walk with him, he informed us that the "first great movement of the fall migration" had started the previous night, so we should see some "good birds" on our walk.

These "good birds" that birders encounter on walks often include small migrating songbirds that can be very difficult to see, like warblers and vireos. Warblers are smaller than sparrows, typically four to five inches in length. Vireos are the same size as warblers, but "usually less active," according to field guides, meaning that they not only flit around less often than warblers but also are even more difficult to see than warblers.[20] Both warblers and vireos are small songbirds that birders in the New York metropolitan area can see when the birds migrate north for breeding and nesting. Nearly every sighting of a warbler on a walk garners a response of "that's a good bird," and sometimes birders fear that the birds are so good, in this case meaning so rare, that it would likely be their only chance to see them. For example, on a walk with Diane, she heard a Blackburnian Warbler and told the group to look out for him, saying, "They're too good to stay around here for very long." The naturalist gaze teaches birders to appreciate the chances they get to see these birds, knowing their migration route brings them through the area only twice a year. Mona said this makes her and other birders appreciate their temporality, knowing it may be their only chance to see them that year, even if birds' agency and ability to fly away can be frustrating: "A lot of times, people who have birded for years and years, they won't even say, 'I just saw a Crow.' To them, that's just a nah, nah bird, or a Mockingbird. But if they spot, during migration season, a Grosbeak, it's 'Oh, a Grosbeak!' and then, as you begin to walk through as a group, you will just follow the lead of all the other birders and hope that you can spot, if you want to see every bird, a lot of times, if you're not in the front group, or if you're not a real quick response person, you may miss a good bird that you hadn't seen in a year, and that gets a little frustrating. You really want to look at that bird." As discussed in previous chapters, birders also appreciate common birds, like American Crows and Northern Mockingbirds. But as those resident birds can be seen any time of the year, the temporality of migrating birds like Grosbeaks makes birders evaluate them as the "good" birds "during migration season," as Mona specified. Through the naturalist gaze, birders appreciate birds that belong in the ecosystem, even if they're just migrating through it. In contrast, as discussed in chapter 2, birders pity vagrant birds that are out of place and are not meant to be migrating through an area. Birders also view certain behaviors of birds that belong in an ecosystem as "good," even if those behaviors are distasteful.

BAD BIRDS, ANIMAL AGENCY, AND HUMAN INTERVENTION

When I asked birders if they had a least favorite bird, the majority of them named European Starlings, House Sparrows, and Brown-headed Cowbirds. On the surface, their answers make intuitive sense—starlings and house sparrows are invasive, nonnative species, and cowbirds are parasitic species, meaning that all three birds endanger other birds and their habitats. Beneath the surface lies an important understanding of animal agency, and the designation as "bad" birds was not as straightforward as it seemed.

Since the naturalist gaze is evaluative and integrative, when birders evaluate certain birds as "bad," they do so with an understanding of the role that humans play in these birds' lives. Birders know that people brought European Starlings and House Sparrows to North America in the nineteenth century, and they recognize that it's not the birds' fault that they now exist here, in North America, as invasive species. Cowbirds, in contrast, are parasitic by nature—they are following instinct when they lay their eggs in other birds' nests. However, even though cowbirds' badness is instinctual, evolutionary behavior, and even though humans created the environmental degradation that allows cowbirds to thrive, cowbirds' parasitism was so distasteful to birders that they failed to use the naturalist gaze and instead evaluated cowbirds as if they were engaging in purposeful, moral choices. In doing so, they attribute animal agency to all three species, even though they blame one species for its instinctual behavior (cowbirds) while excusing it in others (sparrows and starlings).

In the remainder of the chapter, I focus on three species of birds that predominate birders' discussion of "bad" birds: European Starlings, House Sparrows, and Brown-headed Cowbirds. The comparison of how birders discuss these three species demonstrates the edge between animal agency and instinct, and it will show how birders take into account the role of humans in creating opportunities for these birds to be "bad."

European Starlings and House Sparrows

Both European Starlings and House (or "English") Sparrows are invasive, nonnative species in North America. They are native to Europe and were purposefully brought to North America. On several walks, and in several interviews, birders told me about the "origin story" of European Starlings. Each version always varied a little bit. One person said that someone brought sixty starlings to Central Park, and another person said it was just a box of starlings. One person said it happened in the 1960s, but another birder corrected them and said it was the early 1900s. One person said they heard the birds were brought over for a wedding. Another person said the original release happened in Prospect Park. This was the stuff of lore. Almost any time we saw a starling on a walk, the leader would launch into his or her version of the starling origin story.

In fact, both European Starlings and House Sparrows were brought over by the American Acclimatization Society in the middle of the nineteenth century. "Acclimatization" meant to "introduce, acclimatize, and domesticate all innocuous animals, birds, fishes, insects, and vegetables, whether useful or ornamental."[21] A *New York Times* write-up of an 1877 meeting of the American Acclimatization Society in New York City details the starlings' release: "Last July the Acclimatization Society freed in the park some starlings and Japanese finches; Mr. John Sutherland had done the same with some English pheasants. It was expected that they would all prosper. Mr. Conklin suggested that renewed and organized efforts should be made to acclimatize the English titmouse, chaffinch, blackbird, robin redbreast, and the skylark—birds which were useful to the farmer and contributed to the beauty of the groves and fields."[22] Eugene Schiefflin, chair of the American Acclimatization Society, wanted to bring over all of the birds that were mentioned in the works of Shakespeare. This initial release of starlings was not successful, so Schiefflin brought over more in 1890 and 1891, and since then they have prospered.

This 1877 American Acclimatization Society report also mentioned House Sparrows: "In 1864, he said, the Commissioners of Central Park set free 50 pairs of English sparrows, and they had multiplied amazingly."[23] Much like the multiple attempts to introduce European Starlings, this was not the first attempt to bring over House Sparrows. The Brooklyn Institute brought eight pairs of House Sparrows from England in the fall of 1850, and when they did not thrive, it brought more in 1852 and 1853.[24] Unlike starlings, House Sparrows did not have literary origins—they were brought over to control insects plaguing farmers' crops.[25]

Now, more than 200 million European Starlings and more than 150 million House Sparrows can be found throughout North America, and they are among the most common birds on the continent. What makes starlings and sparrows invasive is their effect on native bird populations. European Starlings compete with cavity-nesting birds such as Bluebirds, Purple Martins, and Woodpeckers, often taking over their nests and expelling the birds already there. Furthermore, starlings create a significant impact on the ecosystem because of their large numbers. They also cause considerable damage to a variety of forms of agriculture: with their large flocks, often numbering in the thousands, they can pull up entire fields of young plants to eat the seeds, and when they gather at feed troughs on animal farms, they contaminate the food and water sources. House Sparrows also compete with native species, especially Bluebirds, for space in nest boxes. Ornithologists at the Cornell Lab explain that their competitive, territorial nature, combined with their prolific breeding (raising up to four broods per season, in comparison with two for Bluebirds), makes them especially dangerous to native species.[26] Because they are invasive, nonnative species, European Starlings and House Sparrows are not protected under the Migratory Bird Treaty Act of 1918, which makes it illegal to hunt, kill, capture, or sell migratory birds. And in 2004, European Starlings accounted for 2,320,086, or 84 percent, of all animals killed by the USDA's Wildlife Services Division.[27] (Sparrows accounted for 15,508, or 5.6 percent, of all animals killed.)

Birders employ the evaluative and concerned aspects of the naturalist gaze when they discuss European Starlings and House Sparrows. The naturalist gaze is evaluative, judging the flora and fauna in an ecosystem on the basis of scientific information. Additionally, the naturalist gaze is concerned about the health and well-being of wildlife in its natural habitat. When birders judge European Starlings and House Sparrows as "bad birds," they do so because they are invasive, nonnative species whose presence and behaviors harm the ecosystem, which includes other, native birds.

The birders I studied evaluated European Starlings and House Sparrows as "bad birds" primarily because of their status as invasive, nonnative species. As Cheryl said in our interview, she assumed that no birders liked them because of their invasiveness: "I don't think any birder likes the House Sparrows and the Starlings, because they're invasive." The birders I interviewed and observed were primarily concerned with these invasive birds' effects on the local ecosystems used by other, native animals, as Ethan explained: "I don't like invasive species. So Starlings, not my friend. I don't like Starlings, I don't like House Sparrows. Perhaps in their native range, they'd be cool, but in terms of being a destructive organism, in a place that they're not supposed to be, it's kind of, it's not very happy to know that such an animal is destroying or modified other native creatures' habitats by outcompeting them for natural resources." Ethan worried about these invasive species outcompeting the native birds for natural resources, which demonstrates the evaluative and concerned aspects of the naturalist gaze. Other birders were also concerned about the invasive species eating the birdseed they put out for native birds in the winter, as Mona explained: "Of course I don't like the English Sparrow, I don't like the Starlings, but I love to listen to Starlings, I love to look at them, I love the color of them. What I don't like about them, the Starlings and the Grackles too, they come to your feeder and they wipe it out and there's none left for the smaller birds. With juggling the feeders, and keeping different feeders filled, the ones they can and can't get on, you can bring about a balance for that, but they still have a way of scurrying away all the other birds." Like other birders who assumed that everyone dislikes Starlings and Sparrows, Mona said "of course" she didn't like those invasive species. But her primary concern was that the invasive species ate the feeder food that she put out for the smaller, native species of birds. Anyone who has ever put out birdseed has likely experienced this phenomenon—dozens and dozens of European Starlings and House Sparrows come and feed first, leaving little for other birds. Because of this, Miranda said that European Starlings give all birds a bad name: "I really don't like starlings. European Starlings don't belong here, they come and they clean me out of all my birdseed at home. They're loud, they kind of give birds a bad name. That's what people think of when they think, when they have a negative perception of birds, it's because of these big flocks of birds, and then it's overwhelming, noise-wise, or motion-wise, and so I feel like they're bad bird ambassadors. They don't represent their fellows very well."

Because European Starlings and House Sparrows are invasive, nonnative species, they do not fit into an ecological niche in North America and do not have natural predators, and thus their numbers skyrocket. Given these factors, both species proliferate to such an extent that their sheer numbers can overwhelm casual observers, backyard birders who put out birdseed, and farmers whose crops and livestock are the reason for the USDA's killing of European Starlings by the millions. Sparrows and starlings consume the insects and grains they find in the wild, but they also gladly help themselves to the birdseed that birders put out for other, native songbirds. Sparrows and starlings are not "invited" to feed at bird feeders, and several brands of birdseed specifically avoid ingredients sparrows and starlings prefer, such as corn and millet, so as to deter the unwanted birds.[28]

Despite all of these problems, sparrows and starlings were not completely bad in the eyes of birders. Birders viewed them as bad because they were invasive and thus harmed the habitat and ecosystem for other birds, but at the same time, birders did not consider sparrows and starlings to be wholly bad because of the birds' lack of agency. Birders noted that it was "not their fault" that sparrows and starlings had come to North America. Sparrows and starlings were, ultimately, not to blame for the precise thing that made them "bad"—their invasiveness and nonnativeness.[29] Ultimately, birders knew the fault lies with humans for bringing sparrows and starlings to North America in the first place. Human agency and human choices, not the birds, created the problem for the local ecosystem.

Birders continually noted that it was "not the birds' fault" that they were invasive, nonnative species, as Barbara did in our interview: "Yeah, the English House Sparrow. Just because it's a nonnative, aggressive species that has caused the decline of a lot of our cavity nesters due to its behavior. And it's not the bird's fault, it's obviously just doing what it does. It's just being a bird. But it does compete with our cavity nesters, like Bluebirds, and Tree Swallows, and Purple Martins, and it's just terrible. It picks the eggs, breaks them, throws them out, kills the babies, and just builds its nest right on top of the whole mess. That's one of my least favorite birds. And it's not the bird's fault, obviously, it's not his fault he's here." Here, the House Sparrow exhibited some agency in competing with other birds, even killing them, but Barbara said it was not the birds' fault, because it was not their fault that they were in North America to begin with. Barbara made a moral judgment on humans for exercising their moral agency and bringing the birds over, and she did not judge the birds' instinctual behavior, even if distasteful and harmful, because of their lack of agency in being brought to the ecosystem in the first place.[30]

Other birders used their naturalist gaze to note that the House Sparrows and European Starlings probably fill ecological niches where they come from, even though they are ecologically problematic over here, as Cynthia explained: "A House Sparrow might be my least favorite bird. They're just vicious. I would never say things like that to students, because a bird is a bird; it's a living creature and it's not that bird's fault that it's here. I'm sure in England they're totally useful and have their niche and fit well into the ecosystem." Cynthia works as an environmental educator with children, so she acknowledged House Sparrows' usefulness to her work in a major city: "I guess they have their purpose in the city too, because they're there, and they're a bird, and they're kind of omnipresent, and they help kids learn." House Sparrows' invasiveness had a silver lining, in that their ubiquity allowed her to use them to teach children about birds. She went on to say, "I'm sure they are great birds where they come from," in England, where they fit into their ecosystem.

Likewise, Tom said he knew it was not the starlings' and house sparrows' fault that they were brought to North America. And because they are not native to the United States, he also knew that European Starlings and House Sparrows are not protected under the Migratory Bird Treaty Act of 1918. Nevertheless, Tom said that they still deserve compassion:

> They didn't bring themselves to this. Just because people brought them to the Northern Hemisphere, it's not their fault. Just because they found the place so successful. . . . They wanted all the birds that Shakespeare ever mentioned to be on the North American continent, so they brought in these things, like the House Sparrows and the Starlings, they just brought them to New York in the mid-1800s and now they're just all over. And in fact, I think federal law protects all the birds except the House Sparrows and the Starlings, they're the only ones you can kill, unless you have a conscience and you say I'm not going to hurt the bird, anyway.

Tom didn't spend much time discussing the "bad" behaviors of starlings and sparrows; instead, he put them into the context of having been brought to North America in the first place. Tom suggested people can use their conscience and their human agency to choose not to harm these birds, and he even talked about how he feeds the birds in his yard, including sparrows and starlings.

Other birders marveled at starlings' and sparrows' determination to survive, even in their nonnative habitats. Trish said that she was impressed with how adaptable sparrows were. She talked about traveling throughout Central and South America and seeing House Sparrows everywhere: "These are incredibly adapted to anything birds. It's pretty impressive. To be perfectly honest, I get tired of seeing them, yes, but you've got to appreciate their tenacity, or whatever it is, that has allowed them to be so successful." Instead of decrying them for existing and persisting outside of their native ecosystems, Trish tried to use their ubiquity as an opportunity to talk about providing good habitats for native birds: "If you're talking to people, if you're leading a bird walk, talking to people about birds, and you're seeing sparrows and pigeons, you could bring the conversation to, you know, how birds have different needs, different habitats, and in fact these birds, in a sense, have been displacing other birds that we used to see here more frequently. And is that just these new birds coming in, are we doing other things like killing off a lot of insects in our environment so that the insect-eating birds don't really have a reason to be hanging around? We could start talking about things like that."

Trish turns the problem of seeing nothing but sparrows and pigeons into an opportunity to talk about what people can do to improve the environment for native birds, and how birds, insects, and people all fit together in a shared ecosystem. Like the other birders, she blames people for harming the environment, which harms native birds. Thus, European Starlings and House Sparrows are invasive, nonnative bird species that put other birds in danger, sometimes even killing them, but birders acknowledge it's not the birds' fault that they are here. Ultimately, humans are to blame for bringing these nonnative species to North America and damaging the ecosystem.

Brown-headed Cowbirds

With "good" birds, birders adhered to an understanding of animal agency that aligns with animal studies definitions—birds express their agency when they engage in self-initiated behavior, which can be borne of instinct. Sometimes these instincts and behaviors are distasteful, like in the case of predation or carrion eaters, but, viewing them through the naturalist gaze, birders still see these birds as good because their actions benefit the ecosystem. Still using the naturalist gaze, birders also evaluated European Starlings and House Sparrows as "bad" because their actions harm the ecosystem, though birders mitigated their judgments since these "bad" birds were invasive, nonnative species on the North American continent only because of humans' actions.

In contrast, it was a native North American bird species that drew the most ire from birders—the Brown-headed Cowbird. Birders dislike Brown-headed Cowbirds because they are "brood parasites," meaning they lay their eggs in other birds' nests, to be hatched and reared by the other birds. Cowbirds don't even know how to build their own nests—they exclusively use the nests of other birds for their eggs. Some other species of birds are also brood parasites, primarily cuckoos, but they typically lay their eggs in the nests of other cuckoo subspecies. Instead of solely viewing Brown-headed Cowbirds' behavior through the naturalist gaze, and instead of seeing them as merely following instinct, when birders evaluated cowbirds' behavior, they used the more traditional sociological definition of agency and imbued these birds' behaviors with a sense of free will and moral choice.

In their sociozoologic scale, Arluke and Sanders called the worst animals "demons," which are often portrayed as predators that harm humans.[31] With "demon" animals, such as Pit Bull dogs, the demonic aspect seems inherent to the animals because of *what they are* rather than *what they do*.[32] Thus, while cowbirds may seem like "demons" on the sociozoologic scale, because they raise the most ire from birders, cowbirds exercise more agency than Arluke and Sanders afford such "demonic" animals. Cowbirds are "bad" birds because of *what they do* (i.e., brood parasitism), in contrast to the starlings and sparrows, which are bad because of what humans *did to them* (i.e., bring them to North America). These examples from the birding world offer us new ways of understanding the relationship among human agency, animal agency, and animal instincts.

Birders particularly dislike cowbirds because they lay their eggs in the nests of songbirds, which are much smaller than cowbirds, and then those tiny birds expend a terrific amount of energy raising the comparatively huge juvenile cowbird. Sometimes the songbirds' own chicks die because they do not get enough food, since the cowbird is that much larger than the songbird chicks. Brown-headed Cowbirds are partly to blame for the decline of endangered species like the Kirtland's Warbler or the Black-capped Vireo. Cowbirds parasitize over 220 species of birds, and therefore their effect on the ecosystem is much larger than that of other brood parasites.[33] In our interview, Nancy and Andy incorporated all of these elements as they explained their disdain for Brown-headed Cowbirds:

NANCY: So, you know that we hate the Cowbird.

ANDY: Brown-headed Cowbird.

EC: Why do you say that?

NANCY: They're a parasitic nesting bird. They don't raise their own young. They'll find a nest, they'll lay their egg in it, hope that the bird whose nest they have laid it in doesn't notice. And that bird, which is usually a smaller songbird—

ANDY: Almost always, yeah.

NANCY: Will hatch that egg, raise the babies, the baby will be larger than the other babies, will terrorize them, and will basically push themselves to the front of the pecking order to be fed more, and so that songbird will lose young based on this one bird laying one egg in that nest.

ANDY: It's, so you'll see little warblers and sparrows feeding these giant black, well, not giant blackbirds, but much bigger—

NANCY: —Like double the size.

ANDY: Yeah, double the size of the babies, following the parent around, so you lose some of the songbirds, the other songbirds, like warblers and sparrows, to this blackbird, cowbird. It's frustrating.

NANCY: And it's birds that are of concern sometimes, that are losing population. And these awful, parasitic birds go in, and the population of this songbird is already dwindling, and they're making it dwindle further, by putting in their egg in there, because they're too lazy to raise their own babies.

ANDY: And they've increased because we've fragmented the environment and the forest tracks, and they look on the edges of the forest, and they have easier access to bird nests, to the native bird nests. So that's part of the reason they thrive.

NANCY: And you can't kill them.

ANDY: They're protected.

NANCY: They're protected. Which I think should change. But that's my own opinion.

Nancy and Andy laid out all of the reasons that birders dislike cowbirds: they are brood parasites whose parasitism puts other birds in danger, especially smaller, endangered songbirds. Further, Nancy and Andy noted that since cowbirds are native species in North America, they are protected under the Migratory Bird Treaty Act. Some states, such as Michigan and Texas, give permits to people to trap cowbirds when they threaten endangered species such as the Kirtland's Warbler, the Golden-cheeked Warbler, and the Black-capped Vireo.[34] Nancy and Andy's negative depictions of the Brown-headed Cowbirds primarily focus on the cowbirds' actions: they are "parasitic," they "don't raise their own young," they "terrorize" other birds, they "push themselves to the front of the pecking order to be fed more," and they are "too lazy to raise their own babies." Birders view these cowbirds as engaging in purposeful actions, using their animal agency, and thus, birders evaluate cowbirds with a moral judgment, in addition to their evaluation through the naturalist gaze. Unlike the distasteful but useful behavior of carrion eaters, these cowbirds' distasteful behaviors harm the ecosystem, which includes other birds.

However, human agency also plays a role in Brown-headed Cowbirds' proliferation and endangerment of other songbirds. Toward the end of that exchange, Andy notes that part of why cowbirds thrive is because people have fragmented forest habitats, which gives cowbirds easier access to other birds' nests. Brown-headed Cowbirds got their name because they followed herds of bison in order to eat the insects kicked up by the bison's hooves. Now, cow-

FIGURE 7. Brown-headed Cowbird (photograph courtesy of Dave Saunders).

birds' preferred habitat includes a wide variety of places: open grasslands, agricultural fields, and even urban and suburban habitats with disturbed soil.[35] People's fragmentation of forested areas for development and expansion of agricultural fields provides even more habitat for cowbirds. Additionally, cowbirds lay their eggs in "edge habitats," or the edge of a forest next to such open areas. Biologists have found that in areas with highly fragmented forests, cowbirds use the entire forest, viewing it as "all edge."[36] This combination of expanded habitats for cowbirds and reduced habitats for forest-dwelling songbirds results in even greater opportunities for cowbirds' brood parasitism. These changes to the North American landscape are entirely the fault of humans, much like the humans who brought European Starlings and House Sparrows to North America. However, in contrast to starlings and sparrows, which catch a break because of their lack of agency in their coming to North America, cowbirds get most of the blame from birders because of their own actions, because of their exerting their animal agency through their brood parasitism.

Cowbirds are such notoriously "bad" birds within the birding community that Ellie felt obligated to say that she disliked cowbirds when I asked her if she had a least favorite bird: "I feel like the answer should be Brown-headed Cowbird." Birders' disdain for cowbirds' parasitism is so widespread and well known that Audubon published an online article answering the question "Is it okay to remove Cowbird eggs from host nests?"[37] (The answer was no, because doing so results in the cowbirds returning to the nest and ransacking it, destroying all of the eggs.) Instead of evaluating cowbirds' behavior through the naturalist gaze, and calling it distasteful only because it harms other birds, birders anthropomorphized cowbirds to be bad parents, as Melinda did: "I don't like the way they do their young. They'll lay that egg in another bird's nest and have nothing to do with it. And that just breaks my heart." Cowbirds' neglectful parenting broke Melinda's heart, and it angered Kay: "I also don't like the cowbirds because of what they do. How they push out those eggs and lay their own. That's insidious. Make your own damn nest. Raise your own kids. You know what I mean? I don't like them for what they do." Melinda and Kay anthropomorphized cowbirds and evaluated their distasteful behavior as if it were not instinct but instead represented a purposeful, moral choice.

Whether anthropomorphized or not, these birders expressed their disdain for the cowbirds' animal agency, or their "capacity for self-willed action."[38] Bees "are a purposeful and relatively autonomous species."[39] Bees sting—that's "what they do." Similarly, one could say that cowbirds are parasitic—that's "what they do." In our interview, Tom took both sociological and animal studies definitions of agency into account. He described humans as having choice and free will, but animals are simply following their instincts: "The cowbird is parasitic, but that's its nature. It didn't teach itself to be that, that's the way it is. People teach themselves to be this or that. Animals are made that way. At least we think they are." Tom emphasized that parasitism is part of the "nature" of cowbirds; that's just "the way it is," not something cowbirds learned or decided to do.

However, ornithologists recently discovered that cowbirds *do* learn to act that way. New biological research shows that young cowbirds sneak out of their hosts' nests at night to mingle with their own kind and, well, learn how to be a cowbird.[40] This research came out after I had completed my interviews, and it complicates birders' evaluations of cowbirds even more. If cowbirds learn to become brood parasites from other cowbirds, does this count as purposeful behavior, since it's learned? Tom somewhat exonerated cowbirds by saying "that's it's nature," but if it is learned, not instinctual, behavior, this could lend even more credence to birders' negative evaluation of cowbirds as "bad" birds.

By taking into account human error, animal instincts, animal agency, and ecosystem needs, birders evaluate which species are "good" and which species are "bad" through the naturalist gaze. Birders excuse the European Starlings and House Sparrows' behavior because it's "not their fault" that they are invasive, nonnative species, since they were introduced to North America by humans. Cowbirds, in contrast, are "bad" because their parasitic behavior is their fault—they exercise their animal agency in ways detrimental to other bird species. The naturalist gaze provides part of birders' negative evaluation of cowbirds, since they endanger other native birds, and since humans' habitat destruction provides more edge habitat for cowbirds. Nevertheless, cowbirds' parasitism is so distasteful that birders also anthropomorphized birds and viewed it as a moral failing.

Part of what makes certain animals "bad" is that people only notice them being "bad." The historian Jennifer Martin explains that the Shark Attack File was created in 1958 with the purpose of understanding which activities or conditions provoke shark attacks.[41] The Shark Attack File lists all of the shark attacks that happened in the United States. Were the sharks exerting agency? Or were they just following instinct? Martin argues that sharks are stripped of their agency because no one studied the conditions under which sharks do *not* attack. The creators of the file have no control data, and therefore they cannot tell whether sharks chose not to attack under the same conditions that previously provoked attacks.

In contrast, the naturalist gaze exhorts birders to watch all birds and their behavior at all times. This means that birders pay attention to birds when they are engaging in their everyday behavior, which may be considered "good," and they are not solely paying attention to the birds when they are engaging in distasteful, unwanted, or other "bad" behavior. The naturalist gaze compels them to be complete in their observations. This holistic gaze distinguishes birders from others who tend to pay attention to, and evaluate, animals only when they engage in "bad" behaviors.

For carnivorous birds of prey, like hawks, birders understand that's "what they do"—they eat smaller animals, including other birds. Ornithologists and birders alike know that Brown-headed Cowbirds *never* make their own nests—they are brood parasites by nature, having developed this habit long ago. We may not know the conditions under which sharks choose not to attack, but we do know that cowbirds never build their own nests. Thus, even though cowbirds don't "choose" to be brood parasites, cowbirds are "bad" because their behavior is so widespread that they have put other birds at risk of becoming extinct. The scale at which cowbirds kill other birds is enough to tip them into the "bad" category.

The naturalist gaze is evaluative, and birders use it to categorize birds as "good" or "bad." The naturalist gaze is also integrative, viewing birds and humans as members of a shared ecosystem. "Good" birds belong in an ecosystem and provide ecosystem benefits. "Bad" birds harm the ecosystem through their actions, though sometimes their behavior can be excused when humans were ultimately to blame. This chapter showed the edge where animal agency and instinct meet.

6 BIRDING AND CITIZEN SCIENCE

ONE OF THE most popular and longest-running citizen science projects in the world is the annual Christmas Bird Count (CBC), sponsored by the National Audubon Society. The CBC was originally proposed by an amateur ornithologist, Frank Chapman. Chapman founded a journal called *Bird-Lore* in 1899, in which he proposed the first Christmas Bird Count as an alternative to the conventional Christmas Day hunts.[1] Traditionally, "sportsmen's journals" encouraged hunters to kill the largest number of birds possible on Christmas Day, with the results published in the journals. Chapman proposed in his journal *Bird-Lore* that birders count the number of birds of different species that they could find in a single day, hoping to capitalize on this same competitive spirit. Chapman said, "It was to the hunters' instinct and spirit of competition to which we appealed and, if we are not mistaken, it is these elemental traits, rather than interest in the science of ornithology, that still animates the census-taker."[2]

In the first CBC, in 1900, 27 participants submitted bird lists; in 1909, more than 200 participants compiled over 150,000 birds.[3] During this time, the count moved from lone census takers to teams of birders. This way, they could cover more ground and stay out for longer periods of time, thus producing more data. In 1934, the National Association of Audubon Societies purchased *Bird-Lore* from Chapman and took over the CBC. The CBC remains the nation's longest-running citizen science project, and on December 17, 2017, I participated alongside 44 people, on 14 teams, covering the entirety of a county, and traveling 64 miles by foot, as we took part in the National Audubon Society's 118th annual CBC.

I had never participated in the CBC before, primarily because its specified range dates (the two weeks before and after Christmas) and, more specifically, the dates when my local Audubon chapters hold their annual CBCs always fall during my end-of-semester grading, just after final exams. This year, I vowed to take part in the CBC, grading be damned (or pushed back one day).

December can be frigid in the New York metropolitan area, and at a meeting of the local Audubon chapter a few days before I participated in its CBC, I asked several CBC veterans if they had any tips for me as a beginner. They had been discussing the early start times, with several groups meeting at 5:30 A.M. to conduct "owling" before the more typical dawn start to a day of birding. Knowing that the day ended at dusk, around 4:30 P.M., I feared the day would be grueling. "Wear layers," Zoe said, "even for your gloves. You can always take layers off if you get too warm." "Bring a thermos of hot tea," Tom said, "that'll keep you going most of the day."

The night before the CBC, I laid out all of my clothes, making sure to include several layers: thermal tights to wear under my jeans, a thermal sweatshirt to wear over two shirts and under a thick hoodie, and my warmest scarf and hat, as well as two pairs of socks and two pairs of gloves, as Zoe recommended. Of course, I'd wear my warmest waterproof boots and coat, and I packed three sets of hand warmers in my backpack. I cleaned my binoculars and packed them, and then I prepared two thermoses for the next day—one for hot coffee, one for hot tea—as well as some sandwiches and granola bars. I had already prepared my vegetarian chili for the post-CBC potluck, to be held in the early evening, once the sun set. I was ready! Or so I thought. I realized that I had little idea what a CBC would look like and how it would compare with a regular day of birding.

Clearly, I feared it would be demanding because of the length of the day spent in the cold. But as far as the birding went, and especially the technical details of a count, I had no idea what it would look like. How would it compare with a regular bird walk, where we keep a species count and then turn that in to eBird? Would everyone become very serious, and no one would joke around, because we were, on that day, officially conducting citizen science?

In this chapter, I show how the naturalist gaze makes birders excellent citizen scientists. The scientific research, field guides, and environmental and wildlife conservation information that inform the naturalist gaze help birders understand the scientific and conservation reasons behind the need for crowdsourced data from citizen scientists. Birders' concern for the health and well-being of wildlife in its natural habitat, another element of the naturalist gaze, compels birders to join citizen science projects to help with wildlife conservation efforts. The instructive facets of the naturalist gaze, through which birders learn how to carefully observe wildlife, help birders collect high-quality data for professional scientists. Finally, the pleasurable aspect of the naturalist gaze—the fact that birding is fun—helps motivate birders to participate in citizen science projects, since they know they will get to observe birds.

However, I also found that birders perceived their contributions to science as less valuable than they were in reality. These misperceptions, I argue, are due to cultural and historical divides between amateur and professional scientists. I show the evolution of ornithology from a practice anyone could engage in to something reserved for professionally trained ornithologists. I also discuss how professional scientists and "amateur ornithologists," or birders, view their participation in citizen science projects. I found that birders share the same skepticism of citizen science as that of professional scientists, because they worry such data will not be useful to professional scientists. If this is the case, why do they participate? How do birders still find satisfaction in participating in citizen science projects? Many citizen science projects employ the exact same activities that birders engage in for leisure, with the only difference being to whom birders report their findings. The symbolic and real benefits of contributing to conservation efforts for birds helped birders find satisfaction in participating in such projects. Some birders also used citizen science as a tool to get students interested in science and wildlife conservation. If professional scientists can better demonstrate the utility of citizen science data to other scientists, and especially to citizen scientists, this will benefit the birders who crowdsource the data, the scientists who use the data, and the birds and other wildlife under study.

CBC AS CITIZEN SCIENCE

Most generally, citizen science describes scientific projects carried out by nonprofessional scientists. Professional scientists typically conceive of the projects and then ask the public for help in gathering the data. This crowdsourcing of data allows for more data to be collected over larger areas and longer periods of time than data collected solely by professional scientists. Audubon's CBC presents a perfect example of a citizen science project.

Before participating in my first CBC, I wondered whether and how bird walks specifically for citizen science counts differed from regular bird walks. I pondered these questions as I drove to the first CBC location, where I met my team at a landfill. Lenny, our team leader, introduced me to Harvey, our other team member. Lenny explained that the landfill used to be better for birding, when it was an open landfill, but the county covered it with grass, which it keeps way too short to provide a good habitat. (Landfills are, in fact, excellent venues for birding.)[4] "We'll walk around the gravel path around the landfill," Lenny said, "and we'll end up at the wastewater treatment plant, where we would normally see a ton of gulls." With that, we started walking and were on our way.

As we walked, we each pointed out various birds and made sure to show them to the other people on our team. We always made sure to include the number, such as "there are two Northern Mockingbirds over there." Sometimes, one of us would ask one of the others, "How many Song Sparrows did you count?" and we would all compare our counts. Once we started the walk, I realized that this CBC walk so far seemed to be like any other bird walk, except with fewer participants. We even stopped to admire certain birds, noting the striking black-and-white feathers of the Hooded Mergansers, or the fight that two Northern Mockingbirds engaged in.

I was just about to ask Lenny what makes this walk different from a regular bird walk when he started to explain how the CBC functions. He told me the county is divided into ten districts, and each district has one or two teams of usually two to four people. "You don't want the teams to be too large," Lenny explained, "otherwise it becomes more of a social thing, and a large,

FIGURE 8. A small team of birders conducting a Christmas Bird Count, with snow still on the ground (photograph by author).

loud group can scare off the birds." He continued, "You don't really need fifteen people to find birds. You just need a few people, because more eyes are better, but there is such a thing as too many people."

He went on to critique the particular local setup for our CBC range: "Our district actually spans the Hudson River, so National Audubon expects us to bird on both sides of the river. But we typically don't do that—our local Audubon group birds in this county, and we don't send people across the river. I even asked the coordinator about it, and I said that you can either have consistency in the number and skill of people who conduct the count over here, or you can expand the count, but we really can't do both. He said that we can just expand the count slowly. So last year, someone in [a town across the river] asked about participating, and we gave him that district, and we were able to include it in our counts." Harvey asked, "Is there a way to separate out those numbers, and say 'these were compiled by [our birding group], and these were conducted by the volunteers from across the river?'" Lenny explained that "National Audubon just compiles the numbers by area, so they wouldn't care about such minutiae in the data."

"So really, other than compiling all of this information collectively, and other than making sure to hit your assigned sites, this really is kind of like a regular bird walk, but you're conducting a census in your area?" I asked. "Yep," Lenny answered. "We just make sure to cover our district, and we count the number of species, and the number of individuals, and then we compile that with the other counters, and turn that in." Lenny did admit that skill matters—you will obviously see more birds if you are a better birder, and you'll be able to identify more birds.

The CBC resembled a regular bird walk, but with less explanation of birds for beginning birders and more cross-checking and verification of species and numbers. Every once in a while, Harvey or I, less experienced birders than Lenny, would misidentify a bird, and Lenny would walk us through the correct identification, demonstrating the informative quality of the naturalist gaze. This practice seemed to be a habit of his from being a longtime walk leader, and it was useful to learn and remember the distinguishing characteristics of several different species. "You asked how to identify a Herring Gull," Lenny said. "Look through the scope at his bill. You can see a red dot and a black dot on the underside of his bill."

The CBC differed from a regular bird walk in that a typical weekend walk with a local Audubon club will visit one location per day, for only a couple of hours in the morning. The CBC is a full-day event, more closely resembling the weekend trips that some Audubon chapters take, where you visit several sites over a couple of days. Once we finished at the landfill, after about an hour, we then got into our cars and drove to a park, where we walked around for forty-five minutes, including a bathroom break and a snack. I left my car at the park and carpooled with Lenny and Harvey for the remainder of the day, as we continued the pattern of driving to a park, walking around, conducting our count, and then driving to another park.

Sometimes we conducted what Lenny called "drive-by birding," where we would drive down a particular road that had a specific feature, such as a certain type of tree or a seawall next to the Hudson River, and we would bird from inside the car. For smaller sites such as those, we didn't get out and walk around—we conducted a census just of the birds we saw from the car. We even found a new species from the car, a House Finch, which we didn't see anywhere else but on that particular road. One time, we stopped the car and drove back to a power line where we saw hundreds of European Starlings perched on the line, and we counted them. Each location got a specific count and time.

Another difference I noticed was that, in comparison with regular bird walks, Lenny and Harvey were more careful not to double count birds. As we walked through one park, Harvey went to the north end and got excited when he saw a pair of Hooded Mergansers. Then he realized that they were probably the same birds that we had seen when we were in the park just north of us, so the two Mergansers didn't go on the list. Each time we saw a group of Canada Geese, we made sure to double-check that they hadn't just flown over from another area in the park. Or, even counting a singular Black-backed Gull, Harvey walked to the other side of the marina to make sure that the other Black-backed Gull was still over there so that we didn't double count the bird.

If there were so many similarities to a regular bird walk, what truly makes the CBC different? The similarities show that birders' careful attention on each of their bird walks, and their submission of their species lists to eBird, can make any bird walk an opportunity for citizen science. The difference between those walks and the CBC lies in the fact that there are dozens of other people, on different teams, all out at the same time, on the same day, being just as careful in their compiling and documentation, who plan to come together to compile this information to submit to the National Audubon Society. This carefully compiled, annual, nationwide census has provided voluminous data for professional scientists to study birds' populations and ranges for over a century.

THE NATURALIST GAZE AND CITIZEN SCIENCE

The basic elements of birding—observing, identifying, and listing—are precisely what citizen scientists provide to citizen science projects, and the naturalist gaze makes birders excellent citizen scientists. Birders learn to accurately identify birds by studying field guides and scientific research on birds, which helps develop the informed element of the naturalist gaze. Dawn, a PhD candidate in environmental studies, told me that her interest in building her naturalist skills led her to learn how to identify flora and fauna by placing them in their ecological contexts. She said, "I like to identify things," and explained that she began building her naturalist skills in high school and college: "I wanted to know what kind of tree that was, and that tree's impact on different species, and the role it played in the greater ecosystem. Birds are so abundant and so vivid and part of that. You're either hearing them or seeing them, most of the time, that identifying them was a really important thing for me to figure out. Why is that Blue Jay in that tree? That tree is a pine tree, so . . ." Here, Dawn describes developing her naturalist gaze—the Blue Jay she observed was not in just any tree; he was in a pine tree. Dawn was able to place the flora and fauna she observed in their ecosystem, and she understood their ecological roles. Nancy, a special projects coordinator with a background in graphic design, described her voracious appetite for learning bird species, which began when she "got the bird books and just started to identify everything I saw, and just tried to know all the birds that I was looking at." Birders with a well-developed naturalist gaze place all of the elements of the environment into context and accurately report what they see, making them adept citizen scientists.

Nearly every person I interviewed participated in some form of citizen science, and the leaders of nearly every bird walk that I observed submitted their species lists to eBird, thus also contributing to citizen science. eBird is an online data repository where birders upload their day's species list, which professional scientists use to monitor birds' ranges. Diane, a regular walk leader and avid eBird user, explained the relationship between birding and citizen science in our interview when she started to define birding: "Watching and appreciating birds, the relationship obviously there is that people then appreciate habitat and are more likely to support environmental protection. But the other key connection is that if birding has the recording component, it's a database to make environmental decisions. And then closing the loop, environmental protection makes for habitats and places for people to enjoy birds, so it comes back around." The everyday lists that birders keep, Diane explained, help people make environmental decisions—as long as the birders submit those lists.

The most popular venue for sharing species lists is eBird. Walk leaders often share their eBird species lists with walk participants after a walk, and birders described eBird as part of their regular birding routine, as Catherine did. First, she told me that the night before a birding trip, she and her husband will set aside their binoculars and their scope, and they'll charge the batteries for their cameras. The morning of, they make sure to eat a hearty breakfast. Then, as soon as they start seeing birds, they begin to keep their species list for the day's walk. She noted the importance of sharing their lists on eBird: "Lenny has been using eBird for the last few years, so he's the record keeper. He's always kept a notebook of what we see . . . he does a count, so if we're just walking around the lake, he'll be writing them down, we'll be taking a count, keep the species list, and then when we get home, he usually enters the data through eBird." As Catherine said, they always kept a notebook of the species they saw, but with the advent of eBird, that recording practice becomes citizen science. Thus, birders learn to accurately identify birds by placing them in their ecological context, and the regular birding practice of keeping a species list further helps birders contribute to citizen science.

DIVIDES BETWEEN PROFESSIONAL AND AMATEUR SCIENTISTS

"Science" used to be socially constructed as a practice anyone could do, especially ornithological science. When birding first began as a hobby in the late nineteenth century, the lines between amateur and professional ornithologists were unclear. Birding emerged as a serious pastime at the end of the nineteenth century because of increasing threats to birds, the emergence of ornithology as a science, the newfound appreciation for leisure time and hobbies, and writing about and drawing birds as part of natural history writing and scientific illustration.[5] Because of these factors, birding became a legitimate leisure pursuit. It combined an interest in the outdoors with an interest in science, and it was now acceptable to enjoy pastimes not directly related to work. The social construction of leisure time had begun (at least for predominantly affluent white men).

During the nineteenth century, people viewed birding as a form of amateur science, and birding became linked with ornithology.[6] In 1900, the amateur ornithologist Frank Chapman founded the CBC. Chapman worked as a clerk at the American Exchange National Bank in New York City, and after he met two ornithologists, his youthful interest in birds was rekindled and he eagerly pursued his new hobby.[7] In 1886, when he was twenty-two years old, Chapman quit his banking job to pursue ornithology full time. With nothing more than his high school degree from Englewood Academy, Chapman first became an assistant to the curator of birds and mammals at the

American Museum of Natural History, and he later became the director of that division. In fact, many famous ornithologists did not have academic training in the discipline at the time. The Cornell Lab of Ornithology did not yet exist, and graduate training in ornithology did not begin in earnest until after World War II. Nonetheless, Chapman's lack of higher education and his ability to climb the ranks at the museum show the blurred boundaries between amateur and professional, especially in ornithology.

In the late nineteenth and early twentieth centuries, the lines between amateur and professional ornithologists became clearer and stronger. Three main factors helped distinguish professionals from amateurs.[8] First, post-Darwinian classification schemas helped professional scientists study relationships between groups of birds and identify subspecies of birds. Second, and related, the idea of scientists having the authority to give order to the natural world, and scientists being accountable to a particular institution, exacerbated these divisions. Finally, biologists, and especially ornithologists, began to change their language. Professional scientists began, post-Darwin, to use technical terms, and these specialized vocabularies helped separate professionals from amateurs.

Around the same time, professional organizations such as the American Ornithologists' Union (AOU) began to emerge. Since only a few of the AOU's members made a living as ornithologists, the organization accepted many "amateur" members, meaning people who did not have ornithology as their occupation.[9] The professional ornithologists in the AOU realized they relied heavily on the amateur members for their specimens and other data, so the amateurs were welcomed into the union.

The rise of graduate degrees in ornithology also created more professional ornithologists and more opportunities for amateurs to participate in citizen science. The Cornell Lab of Ornithology started as a graduate program in ornithology at Cornell University and became a self-supporting research institution after World War II.[10] In 1962, it began its own magazine, *Living Bird*, and in the 1980s, it started its own citizen science projects.

The Cornell Lab sponsors several citizen science projects, including Project FeederWatch, Yard Map, NestWatch, and the Great Backyard Bird Count. Its House Finch Disease Survey was the first citizen science project designed to track the spread of infectious diseases in a wildlife population.[11] eBird, a joint project between the Cornell Lab and the National Audubon Society, was launched in 2002 as a way of using the internet to gather more data. eBird is a website and smartphone app where birders upload their species lists from specific bird walks. In this way, everyday walks that birders take contribute to citizen science projects. The Cornell Lab now hosts over 600 citizen science projects related to a variety of environmental conservation projects.[12]

During the twentieth century, dozens of other universities began graduate programs for ornithology. As ornithology became professionalized and credentialed, it maintained its links to amateur ornithology through citizen science projects. At the same time, these simultaneous trajectories helped create divides between amateur and professional ornithologists. These divides are not clear-cut, though, as most professional ornithologists are also hobbyist birders. Scott Weidensaul, an award-winning science writer for the Audubon Society, writes: "It's hard to find an academically trained ornithologist with a string of initials after his or her name who isn't also an avid birder. I don't know many structural engineers who devote their free time to visiting highway overpasses for fun, but there is something about birds that makes even those whose nine-to-five jobs are ornithological pick up their binoculars as soon as the workday is finished. That's because in almost every case, the job followed the passion, not the other way around. Call it all 'bird study,' and forget the distinctions."[13] These newly drawn lines between amateur and professional do not clarify; rather, they continue to blur. We see this blurring of citizen and scientist as well in hobbies such as beekeeping, where the sociologists Lisa Jean Moore and Mary Kosut write that "scientific beekeepers" seek to conserve bees through empirical observation, intervention, and technology.[14] Although amateur beekeepers do not necessarily share their data with professional entomologists, they use the work of entomologists to inform their hobby.

As noted earlier, citizen science describes scientific projects carried out by nonprofessional scientists. Quite often, scientific researchers and those studying scientists conceive of "citizens" and "scientists" as ideal types.[15] In this ideal typical world, professional scientists hold the professional judgment necessary to conceive of worthy scientific projects, and they do so under wholly objective conditions. Then, they ask the general public, who may hold technical expertise but who lack professional judgment, for help collecting the data. The professional scientists then analyze these crowdsourced data. However, the world of citizen science does not necessarily conform to these ideal typical traits. Sometimes, citizen scientists challenge existing projects, such as AIDS activists who challenged the clinical trials for AIDS research.[16] Or, citizen scientists conceive of entire projects, such as the DIYbio (Do It Yourself biology) projects conceived by citizens who created projects to study the genetics behind their family's hereditary disease.[17] Birders defy these ideal typical scenarios as well, with professional biologists working alongside carpenters on the exact same project. Both people, one a professional scientist and one not, bring the same necessary skills to the table.

DIVIDES BETWEEN WORK AND LEISURE

If the CBC resembles a regular bird walk, and if both CBCs and regular bird walks resemble the work that field biologists do, what differentiates birding from field biology? Very little, other than the profession of the person conducting the monitoring. And even then, the lines are not clear, since professional biologists and ornithologists are often birders who also monitor birds for fun, as part of their birding as a hobby. Birding and citizen science blur the lines between work and leisure.

Earlier, I said that most of the birders I interviewed told me they participated in citizen science projects. The only people I interviewed who said they did not participate in citizen science were those who participated in citizen science as part of their job—environmental educators or people who worked for various Audubon chapters who participated in citizen science projects as part of their job. They did not consider themselves as participating in citizen science, because they did not *also* do so in their spare time. For example, Cynthia claimed she did not conduct citizen science, because she conducted citizen science projects "only" at work—she did not also conduct them in her free time outside of work: "I personally am embarrassed to say I don't have my own eBird account. So, I guess I don't really do my own citizen science and that's interesting. I should. For some reason it's like, 'Well I do it at work, that's like work.' I should probably re-evaluate my feelings on that. I'm a scientist at heart, and I love that, so I really should." Even though Cynthia regularly conducts citizen science projects at work, she chastised herself for not also conducting citizen science projects on her own, in her leisure time.

In the inverse, I also found that birders with doctorates in biology or ornithology who did not get to participate in citizen science projects in their biology-related job did so in their spare time, even if it mirrored their job.

For example, Lenny said that when he moved to New York, he worked for an environmental engineering firm as a biostatistician, and he wanted to "get out into the field and do things," so he started doing field surveys with a local naturalist group. This practice got Lenny interested in birding, because he started seeing the connections between birds and their environment: "I had always been kind of intimidated by learning the birds, especially the little sparrows, they all look alike. But as I started playing with the database and was able to filter things out and see what was showing up when and where, and what things were abundant and what things weren't, it started to click. And we started getting interested, going to the Audubon meetings, and I think what really kicked us into gear was doing the Breeding Bird Survey." Lenny's job as a biostatistician, where he was, as he put it, "chained to a computer pretty much all day long," exemplifies the sociologist C. Wright Mills's critique of white-collar jobs as meaningless.[18] Mills and Hans Gerth later argued that modern work should move to resemble craft professions and what they called the "ethic of craftsmanship," in which workers have more autonomy and creativity in their jobs, which make their jobs more personally meaningful and intrinsically motivating.[19] For craft workers, their work bleeds into their leisure time because of this enjoyment. Likewise, this blurring of work and leisure occurs for ornithologists and biologists who are also birders.

Allen, who works as a wildlife biologist, exemplified this blurring when he described his experience with citizen science at his job and in his leisure time: "I do [citizen science] with my job. We do surveys for different animals. We set out live traps. We do photo, wildlife cameras for the traps. We've got in wetlands, we have water level data loggers, and we do all sorts of thing. But that's my job, so yes. And then with Audubon, we do the Christmas Bird Counts and that kind of thing, the citizen science kind of stuff. I think I'm fortunate that I get to do it while I'm working and during my playtime. I get the opportunity to do that. Whenever I want to, basically." Allen is happy that he "gets to" do citizen science at work and in his free time. Lenny wished he got to do citizen science at work, but instead he was "chained to a computer" conducting biostatistics. All birders thought citizen science was fun, and they enjoy doing citizen science, whether it's part of their job or not. Here, birding and citizen science exemplify the inverse of what the sociologist Jeff Kidder found about bike messengers and edgework.[20] In his ethnography of bike messengers, Kidder found that messengers were willing to take on difficult, dangerous, and poorly paid jobs, such as messaging, simply because they love riding their bikes around the city. Bike messengers bring the thrill-seeking aspect of leisure pursuits and edgework into the job of messaging. Messengers make an unpleasant job fun and exciting, because they turn it into a game, a race, a thrill. In contrast, birders bring what many might consider to be a boring job—the mundanity of serious biological science—into their leisure time. They turn their hobby into work.

The scientists who developed eBird fully recognized this element of birders' hobby and capitalized on it to make birding even more useful for their citizen science projects. The ornithologist Brian Sullivan and his colleagues at the Cornell Lab of Ornithology write, "Many projects have asked the question, 'what can birders do for science?', but none have asked, 'how can we build a useful resource for birders while also engaging them in science?'"[21] eBird began as a project with citizen scientists in mind—how can ornithologists capitalize on the behaviors that birders already engage in, and use them for the ends of citizen science? Sullivan and his colleagues note that other citizen science projects typically begin with a scientific question, designed by professional scientists, who then enlist citizen scientists to help with data collection. Even though eBird flips that script and started with the idea of using birders' behaviors as the basis for its data collection tool, birders' hobby of observing, identifying, and listing the bird species they see still resembles the mundane work of field biology.

While nonbirders might look at what birders do and find it boring, birders found it fun and gratifying. Tom said he was "more than happy" to participate in citizen science as a field observer, but he wanted no part in the aspects of the job that he found boring—the analysis of the findings: "There's certain people that go out and get the data, and there's people that analyze data. I'm not part of that group that analyzes it. I just see it for myself. I'm more than happy to report to these people what I see. But I don't want their job. If they don't have the time to do what I do, well, I have the time to do it. I don't mind at all, reporting these facts that I see to you. But I don't want that part of it." Medical doctors delegate grunt work to nurses to maintain their professional boundaries.[22] Nurses perform the lower-status tasks, whereas doctors perform the higher-status tasks, so as to preserve their higher professional status. The fact that ornithologists themselves participate in the fieldwork—and see it as fun, even—makes this relationship seem more symbiotic and egalitarian than traditional hierarchized professions. All the birders want, in return, is a bit of recognition of their efforts, as Tom said at an Audubon meeting one evening. Speaking with a small group of other birders, he described a recent bird count he did for a local park, and said, "I donated my time, and I didn't even get thanked for it. I don't mind doing the work, because I like going out in the woods and looking at birds, but at least I should get thanked for it!" Here we see the fine line between work and play, between keeping a species list for fun and keeping one for citizen science. It is likely that Tom would have been doing the exact same work on a regular walk where he kept a species list, but here the difference was that he was commissioned to do so by the state park. When the exact same habit becomes work, the expectations for recognition change.

SKEPTICISM FROM PROFESSIONAL SCIENTISTS

Earlier, I described how the social construction of ornithological science changed from something anyone could do in the nineteenth century to something exclusive to trained, professional scientists in the twentieth century. Now, birding contributes more to citizen science than any other hobby or sector of society and has done so for years. Despite this long history of birders contributing to citizen science, contemporary scientists are skeptical of the use of citizen science data—from birding or elsewhere—in their projects. I explore this skepticism through existing literature on professional scientists.

Sometimes professional scientists ignore lay knowledge because of arrogance. Scientists ignored Cumbrian sheep farmers' knowledge of their own lands when studying the effects of Chernobyl on the soil.[23] Other times, lay experts successfully embed themselves into scientific groups. For example, in his study of lay expertise in AIDS research, the sociologist Steven Epstein called this phenomenon "scientific credibility," or "the capacity of claims makers to enroll supporters behind their claims, to legitimate their arguments as authoritative knowledge, and to present themselves as the sort of people who can give voice to science."[24] Citizen science projects differ from these examples in that they are commissioned by professional scientists, with an explicit goal of crowdsourcing information from nonprofessional scientists. Citizen science projects related to birding typically have both conservation and investigation goals, with an eye toward promoting stewardship and wildlife conservation.[25] Even when professional scientists commission citizen science projects, scientists worry about the reliability, validity, or quality of the data collected by amateurs.

The CBC and eBird are two of the largest and most extensive citizen science projects that crowdsource data for professional scientists through large numbers of amateur participants.[26] While the number of crowdsourcing tools for data collection by citizen science proliferates, researchers hesitate to use these sources. The sociologists of science Hauke Riesch and Clive Potter interviewed professional scientists about their experiences with citizen science. They found that professional scientists worried about data quality, and their negative perceptions about such citizen science were driven in part by the need to persuade scientific peers, including peer reviewers, about the effectiveness of such data.[27] Similarly, the computer scientist Edith Law and her colleagues interviewed scholars from a variety of fields and found that their apprehensions about crowdsourcing data stemmed from a concern about the quality of data and the need for further verification of any data that came from crowdsourcing or citizen science.[28] Perceptions about data quality did not come from actual experiences with citizen science, however—the study by Law and her colleagues asked only about whether and how researchers might *potentially* use crowdsourced data.

Studying scientists who have already used citizen science data, Jennifer Shirk, a conservation biologist and director of the Citizen Science Association, found that scientists had to negotiate the perceptions of their research among their peers—they had to prove to their peers that they were doing "real science" and not just public service projects.[29] Shirk's findings demonstrate the sociologist of science Naomi Aronson's argument that the primary claims-making activity of science occurs between scientists.[30] Scientists must primarily prove themselves to other scientists, which requires passing peer review, getting published, and then being cited by other scientists. Ironically, Aronson found that popularizing science represents such a low-status activity that only high-status scientists can afford to engage in the practice. The biologist Hillary Burgess and her colleagues found that biodiversity scientists held a bias toward data collected by professional scientists, and they were concerned about inconsistency in the quality of data from citizen science projects.[31] But at the same time, they found that scientists who worked with birds were more likely to use citizen science data than other researchers, because birds are "particularly well known subjects of citizen science."[32]

Researchers who had never used citizen science data were most likely to be skeptical of data collected by citizen scientists.

Scientists working on the International Union for the Conservation of Nature (IUCN) Red List of threatened species use data from citizen science projects such as eBird, BirdTrack, and xeno-canto to map species distributions. But when it comes to defining extinction risk, they say, "It is also fundamental to respect the rigorous system for assessing extinction risk for the Red List."[33] Census projects do not fulfill the same criteria the IUCN require to accurately assess the area of suitable habitat nor the extent of occurrence, they argue, and thus the IUCN needs to maintain consistency in its measurements. Other IUCN researchers argue that Participatory Monitoring and Management Partnerships might be a better approach than citizen science, since they allow for more community-based knowledge creation and conservation management of local lands.[34]

Most citizen science projects employ a variety of techniques to ensure reliability and validity.[35] The zoologist Raoul Mulder and his colleagues studied two specific case studies of citizen science projects involving the reporting of tagged animals, and they found that reporting error was minimized by careful tag design and clear instructions to participants.[36] Moreover, in assessing the quality of the data from these two citizen science projects, they found that the data collected from such reports were voluminous, unique, and generally reliable. This combination of skepticism and a lack of experience using citizen science data demonstrates how professions preserve professional autonomy by drawing boundaries toward outsiders.[37] It shows that scientists may be more concerned with maintaining divisions between professionals and amateurs than they are with scientific rigor.

To be clear, professional scientists extensively use data provided by citizen scientists. Data from the CBC have been used to inform the National Audubon Society's Common Birds in Decline Report and its Climate Change Report, by showing how birds' populations and ranges have changed throughout this time. The Environmental Protection Agency, the North American Bird Conservation Initiative, and the U.S. Fish & Wildlife Service have also used Audubon's CBC data in their reports on climate change and other bird-related reports. Scientists at the Cornell Lab of Ornithology feature citations for hundreds of peer-reviewed publications that use data gathered by citizen scientists on the websites for the CBC and other citizen science projects.[38]

SKEPTICISM FROM BIRDERS

The studies cited earlier interviewed only professional scientists, and studies on amateurs' participation in citizen science projects only explore how to encourage people to participate or the ethics of participation. They do not ask participants what they think about the data they themselves collect. These studies explore "scientific credibility" from the point of view of professional scientists but not from the point of view of those providing data for these citizen science projects.[39] I wanted to know more about the experience of the birders who regularly and eagerly participate in citizen science projects. In this section, when quoting birders, I refer to the educational background of the birders I interviewed, because here such information is relevant—I found different evaluations of citizen science from interviewees with and without an educational or occupational background in the natural sciences. Ten of my thirty interviewees earned master's degrees, and five more held doctoral degrees. Fully half of my interviewees held a degree in the natural sciences, and six of them worked in or retired from a natural science field.

Most of the birders I interviewed participated in citizen science projects, and many of them said they were skeptical of the utility of the data from citizen science projects because they—the birders—were not "real" scientists. This skepticism is somewhat surprising, since birders have developed a culture of honesty in accurately reporting their species lists. This culture of honesty is demonstrated on birding walks, when leaders double-check their species lists as they go. For example, on one walk, the leader said that she had counted sixty Yellow-rumped Warblers thus far on our trip. She then asked the group, "Is that a suspicious number?" As Miranda (MS in biological systems engineering, director of a local Audubon chapter) said:

> You always have to worry about your data when you're doing research. But at the same rate, I tend to think that birders are honest people. They're kind of a self-regulating group. Like if someone calls a bird that it clearly isn't, someone is always like, "Wait, why do you think it's that?" Things like eBird, if you have an unusual bird, they'll come back to you for more information. I think that there's a lot of checks built into the community itself, but also how the data is collected to prevent things like that. And like I said, I think the community itself is very self-policing. Like the World Series of Birding. There are no judges that

go out with each team. I don't think that the teams are lying about what they're seeing.

This culture of honesty exists in friendly competitions like the World Series of Birding, where the rules stipulate that at least two team members must observe a bird for it to go on the team's species list. More importantly for citizen science, this culture of honesty also guides birders' practices when producing species lists, either for regular walks or in citizen science projects. I observed birders making certain not to double count birds when producing census lists for citizen science projects, and double-checking their counts with other birders to ensure the most accurate counts possible. Key to this honesty is the fact that birders know that lying won't help birds—how could an inaccurate species list help professional scientists better understand bird ranges? The purpose of submitting species lists to citizen science projects is to help with wildlife conservation, and so birders were also intrinsically motivated to produce the most accurate lists possible. This honesty built from a knowledge about the purpose of species lists relates to the informed element of the naturalist gaze.

In my interviews, I found that birders constructed symbolic boundaries between themselves and professional scientists, and that this distinction depended on the science background of the birders themselves. The science scholar Gwen Ottinger argued that boundaries between scientists and non-scientists came from the use of scientific instruments, but that explanation doesn't describe birders who use the exact same instruments as ornithologists: binoculars and scopes.[40] Here, I found the boundary came from birders without a natural science background contrasting themselves with "experts," which included professional scientists. In an interview with Helen, who held a bachelor's degree in business and did not work in a natural science field, she said that she has banded birds and helped with bird surveys: "I assist with bird surveys, I assist the biologists, I scribe for them. I also do bird surveys for them." Helen asked me if I knew what a bird survey was, and she explained, "It's like when you're going in an inlet and recording which species you see and how many of each species. That's what a bird survey is." I asked if that was like what people do when they report their species lists on eBird, and Helen said, "Perhaps. But eBird is just, eBird is not science—eBird is not, I don't think eBird is used by biologists. But in a way it's similar, because you're recording numbers and species." I asked her about the bird surveys she conducts and whether she does them for biologists. She said:

> Yeah, for the coastal biologists. Maintain postings, coordinate volunteers. I do a nest census, which is where you actually go into a nesting colony and you count every nest, if you have any questions about that. You actually go with two other biologists—I'm not a biologist, I'm an HR-sociology-business blend—but you actually carry it like a dowel rod, and you're making a scrape in the sand, and the person next to you is making a scrape in the sand, and is counting the nests between her scrape and my scrape, and at the end we tally, you count every nest. And just for your information, at the south end of [our] beach, this nesting season, we have 232 Least Tern nests, and 175 Black Skimmer nests. And we've counted them. And 4 Oystercatchers, 12 Common Terns. Those are big numbers. It's really a successful nesting colony.

Helen socially constructed science as data collected under the supervision of, and used by, professional scientists. She also noted that she was not a biologist, but she quickly and easily described how to do a nest census and exactly how many nests they had on the beach. Helen said she was skeptical of eBird data but not of the nest census carried out under the supervision of professional biologists. Helen's valuing of scientific authority, and devaluing her own expertise, mirrors scientists' own professional boundaries: doctors or scientists will delegate low-status work to lower-status employees in order to maintain their higher status, or they try to maintain their turf by constructing boundaries against other, related, professions.[41] Here, Helen did the same thing, but in the inverse: she gave her own expertise less value simply because she was not a professional scientist.

Other birders without a natural science background also deferred to scientists' expertise. Dawn (PhD candidate in environmental studies, director of education for a local Audubon chapter) noted that the monitoring at her local Audubon's Important Bird Areas was conducted by biologists rather than amateurs: "monitoring, through citizen science, we do a lot. We also have biologists that do a lot of bird monitoring at our important bird areas and things like that as well."

The birders I interviewed who had a natural science background compared citizen science projects such as the CBC to "real" science surveys, which are conducted under the supervision of professional scientists and which adhere to more stringent scientific protocols. Despite these birders' views that the CBC differed from professional scientific surveys, Joy (MS in horticultural science, retired board member) argued that the CBC data have still been useful for understanding bird ranges:

> A lot of people want to think that, or want to say, well, the Christmas Bird Count doesn't mean anything, because there's no control, there's no true scientific database, but if you go to the same area, year after year after year, even if it's a different person, but you're in the same area, you can certainly see trend lines of where birds are moving. And a good example of that is cardinals, because I think cardinals have moved farther north than they've ever been, and that's been picked up on Christmas Bird Counts, just from people going out and observing them. While it wouldn't pass the test of, you know, maybe a drug study, it certainly passed the test of observation and recognition of what you're seeing.

Joy was playing a bit of the devil's advocate and likely knows from participating in multiple CBCs that the CBC has stringent participation protocols that it outlines on its site, thus ensuring reliability and validity.[42] She also trusted that the CBC provides real results for understanding birds' ranges.

Trish (PhD in microbiology, retired microbiology professor) contrasted the data collection by citizen science with data collected by professional scientists: "Well, considering the fact that scientists do use citizen science data, and I think they realize that it's not the same as professionals going out and doing in-depth surveys, but birding by the general population has provided a lot of data that scientists are now being able to use for looking at trends, what birds are in trouble, what birds are not in trouble, where birds are moving, that sort of thing. So that's all good scientific data." Andy (BS in natural resources management, program manager) also believes in the utility of citizen science data: "I tell people, they really do look at that information when you send it in, and they really do analyze it, and it really does set conservation goals and things like that." While Trish acknowledged that citizen science data differ from "in-depth surveys" conducted by professional scientists, she and other birders with a background in natural science noted that scientists use citizen science data.

The skepticism from birders does not come from laypeople doubting the veracity of science, as the sociologist of science Harry Collins posits, nor does it come from laypeople not understanding science, as the environmental scholar Brian Wynne critiques.[43] Here, the divide comes from birders without a natural science background holding immense respect for scientists as "experts." This has implications for the role of professional scientists in communicating the importance of citizen science, which I discuss further in the following section.

BIRDERS VALUE CITIZEN SCIENCE

Despite their skepticism, all of the birders I interviewed still highly valued citizen science projects. The most common citizen science projects conducted by these birders included the CBC, Project FeederWatch, and the Great Backyard Bird Count, organized by the National Audubon Society and the Cornell Lab of Ornithology. Some birders said they participated in up to five different CBCs every year. Several birders participated in other projects, such as local bird counts for local organizations. Birders valued these projects, and they saw their participation in them as useful and helpful. Even those who expressed some skepticism about the data collected through citizen science projects valued such projects.

The CBC and Project FeederWatch mimic the species lists that birders keep on any other walk—how many birds of which species they saw, on a specific day, and in a specific location. Some citizen science projects in which these birders participated differed from a regular bird walk, though. For example, Joy (MS in horticultural science, retired board member) participated in a project monitoring banded birds: "I just this spring started participating in a project through the Smithsonian. It's a project to look at, the birds were banded in my back yard, and over the next three to four years, look at what you're seeing, if you're seeing them come back, so that they can determine how long some birds return or stay in the same location." Joy signed up for a multiyear citizen science project to monitor banded birds in her yard, showing that she values citizen science to such a great extent that she is willing to donate years of her time to this one project. These citizen science projects mirror the lists birders keep on their regular walks, but they also mirror the practice of natural history, the precursor to biological and earth sciences.[44] Listing birds by species was the original way that natural historians explored and cataloged the world in the eighteenth and early nineteenth centuries.

Now, amateur birders practice the same thing as professional naturalists of the past had done.

Data collection remains the primary objective of the CBC, but personal satisfaction also mattered to participants.[45] While some birders harbored some skepticism about the reliability or validity of the data collected from citizen science projects, all of the birders I interviewed still found personal satisfaction in participating in citizen science projects, and they still found them important for helping birds, as Vivian (MS in management, research operations manager in trademark research) noted: "I like the challenge of the citizen science stuff, but I also, I feel good about contributing. I feel like birding gives me so much, it's a way to give a little bit back, and hopefully make a little bit of a difference. It's unfortunate that a lot of times your news is kind of sad. Things are in decline, or land is less available, all this stuff about climate change. It's not all good news, so that's a little tough. But you've got to figure, even if it ends up being bad news, that information is important to hopefully try to reverse some of that later."

Several of the birders I interviewed used citizen science projects as a way to engage children and students in learning more about science and about birds. Dawn said that she finds citizen science projects to be much more rewarding for students than just learning about the scientific method in a classroom. She said she has them participate in eBird, the Great Backyard Bird Count, and Project FeederWatch: "I do a lot of that with students, and engaging them in observing and collecting data, and then going towards something much bigger and greater than them as an individual, or their class, is really powerful. I mean, a lot of times students have to go through the motions of an activity, or here, do something with the scientific method, but if you can actually apply that to something real and something useful, it's much more impactful." Dawn said that the students she works with find more utility in participating in a citizen science project where their work goes toward the greater good, rather than having students just "go through the motions" for a classroom activity.

For younger students, birds and citizen science provide a venue for getting them interested in conservation issues. Cynthia said she uses citizen science in her teaching of all ages of students now: "Something that we have tried to do is incorporate citizen science into, at first it was with our high school programs. We wanted older kids to be taking conservation action, and citizen science is definitely that, and it very easily fits into their curriculum, taking data and watching how populations change, comparing two different sites and what you see there, and why that is. And I've started actually doing citizen science with kids as young as second grade, using eBird, and teaching them how to use binoculars, and why that's so important." Dawn and Cynthia work as youth outreach coordinators, so engaging students is part of their job. Most of the birders I interviewed did not engage youth in this way, though. They simply saw the value in citizen science projects for helping birds. Dawn explained her broader admiration for citizen science in this way:

> I think that by studying birds and inputting that observation, we're able to learn a ton of things. Through people, just average individuals, laypeople observing and recording bird observations, scientists and researchers have learned a ton of things. They've learned population shifts, and changes in populations, and disease spread, and hybridization, and just amazing amounts of information, that if you just relied on scientists to collect that data, they wouldn't find, it wouldn't be as powerful, because it would take a ton of researchers to be able to have that many eyes in the field. Citizen science for multiple communities is really important, but for bird research, is really important.

Fundamentally, these birders understood that their contributions were valuable to professional scientists, and to the birds. But the underlying skepticism of many birders has implications for professional scientists.

Birding as "amateur ornithology" has many links to professional science and ornithology. As ornithology began to professionalize, it maintained links to the amateurs that provided much of their data. In the twentieth and twenty-first centuries, ornithology and wildlife conservation science moved from being open to anyone to being constructed as the domain of professionalized and credentialed experts. Existing literature explores citizen science only from the perspective of professional scientists, who worry about the validity and reliability of data collected by nonscientists. This chapter shows that birders without a natural science background also share this skepticism of data collected by nonscientists. Birders want citizen science data to be useful for scientists, which means the data must be valid and reliable. But studies show that the data collected by citizen scientists are valid and reliable. This disconnect between lay and expert understandings of the uses of citizen science data has implications for wildlife conservation. For citizen science to be more useful—and believed to be more useful—for wildlife conservation, professional scientists who use citizen science data need to be more vocal about the utility, validity, and reliability of such data. This will help assuage the fears of both professional scientists and the citizen scientists gathering and using such data.

The naturalist gaze provides birders with the skills and motivation to participate in citizen science. Birders know how to accurately observe, identify, and list the species of birds that they see on their walks, and they want to share this information with professional scientists to assist in wildlife conservation efforts. Birders see citizen science as part and parcel of birding, and thus birders are the most active citizen scientists among naturalists. In the next and final chapter, I explore the relationship between birding and wildlife conservation in more detail.

7 BIRDING AS A CONSERVATION MOVEMENT

THE NATURALIST GAZE culminates in birders making conservation efforts to help birds. The naturalist gaze isn't just about looking at birds; it's about changing the world. Once birders develop the naturalist gaze, they engage in a variety of lifestyle changes and conservation advocacy to improve the environment for birds. Birders know that observing and appreciating birds comes first, as Trish explained, "There's always that core appreciation and understanding that if I love birds, and I want to go out and look at birds, the environment has got to be protected, so that I can continue to do this activity." Looking at birds through the naturalist gaze leads to taking actions to benefit birds.

In this chapter, I show how the naturalist gaze moves birders to action. First, birders' careful attention to birds, combined with their citizen science skills, helps them see birds as environmental indicators. Then, birders turn the naturalist gaze on themselves by greening their own lifestyles. Having improved their own lives, birders then turn the naturalist gaze to institutions, encouraging them to make changes to benefit birds, especially birds' habitats. Birders then try to develop the naturalist gaze in others so that they will "create more conservationists," as they put it. In the final stage of their efforts, birders engage in more traditional forms of environmental and wildlife conservation advocacy.

SEEING BIRDS AS ENVIRONMENTAL INDICATORS

The concept of an ecosystem is something we take for granted now, but it was not until the 1950s, when the biologist Eugene Odum began publishing on ecosystem ecology, that the public began to understand the concept as we now know it. An ecosystem describes the system of living organisms and nonliving components of the environment (i.e., air, sun, water, nitrogen, soil) that are linked through the creation and consumption of nutrients and energy. Such a system includes networks and interactions among living organisms, as well as among organisms and the environment. Ecosystems are dynamic, and ecosystem change can come from internal and external factors.

The naturalist gaze helps birders practically understand ecosystems and how to make changes to improve them. As birders develop their naturalist gaze, as they learn to pay careful attention to the natural world, and as they record data in the natural world through their citizen science efforts, they move from looking at birds to taking action to help birds. Seeing birds and people through the integrative lens of the naturalist gaze helps birders view all flora and fauna as interconnected elements in a shared ecosystem. All of the birders I interviewed recognized this environmental interconnectedness of people and birds. As noted in the previous chapter on citizen science, half of my interviewees held a degree in the natural sciences, but even my interviewees who did not also saw birds and people as part of a larger ecosystem. This means birders see the effects people can have on birds and the environment, both positive and negative. This also means birders understand birds as "indicator" or "sentinel" species, meaning they indicate or demonstrate changes in the environment in a variety of ways.

Birders see any detrimental or beneficial effects on the environment as affecting all of us, as Mona said, "By helping birds, they are helping themselves, and mammals, and fish, and every species. You are just helping the entire earth, just starting with a few actions to maybe help the birds. But it's going to benefit everything and everybody living." This general view that we are all connected means that birders understand that environmental actions taken to benefit birds will also benefit other species on the planet, including humans.

Many birders also see the specific ways that particular species are linked, such as fish and Bald Eagles. Chloe told me that "birds are a great way to think about ecosystems, and to understand the ecology of an area." She then gave an example: "If you see Bald Eagles are nesting on the ranch, and you see that, you know there's something happening, they're able to have a sustainable fish population, and so you can make that correlation that it's good, things are good here right now, because of this. They're a good lens on how things are going." Here, Chloe describes birds as an indicator species—birds indicate that things are going well in the environment, if they have enough fish to eat and thrive. This means the water is clean enough to sustain the fish population, which, in turn, sustains the population of fish-eating birds. This acknowledgment that some birds need fish and other birds need other food sources is another element of birders' understanding of the variety of ecosystem needs that various birds hold, as Helen explained: "If you have an appreciation for birds, you have an appreciation for their habitat, and what they need, in my opinion. Because they need bugs. They need flowering plants. It all goes together." Birds need various elements in their habitat—Bald Eagles need fish, which need clean water; some birds need bugs, which need flowering plants.

At the same time, birders recognize that various elements of the ecosystem also need birds in order to thrive. We need birds to pollinate our plants and to distribute seeds throughout the environment, as Joy explained: "They play a tremendous role in the value of the health of our environment. Because they're pollinators, they're insect-eaters, they're seed-eaters, they re-seed things all over the world. If we didn't have them, we wouldn't have the environments that we have across the world—across the country, for sure, but across the world, either." Joy went on to describe how people overlook birds' role in the environment, when people say, "'It's just a bird.' Well, it *is* just a bird, but it also has a big value, and there's another, more emotional value, and that's just that they're fun to listen to!" Joy says birds represent more than we realize; birds provide ecosystem benefits and intangible benefits such as bringing us happiness, which makes up part of the pleasurable element of the naturalist gaze.

Miranda says she wishes people would monetize those ecosystem benefits so that birders could use those arguments for conservation: "They provide ecosystem benefits to us. . . . They eat pests, and bugs that we don't like, and they pollinate things, and they spread the seed of all kinds of plants, so there's intangible ecosystem benefits that could be monetized. And I wish someone would monetize specifically birds' roles in providing those benefits to us, so that we could use that as an argument for conservation." Monetizing environmental benefits is becoming a trend—newly planted trees sport signs indicating the dollar amount of their "services" for cleaning the air. Many environmentalists critique this capitalistic monetization of ecosystem benefits, while also wondering if this tactic is the only way to get people to care about the environment.

Birds' pollination or seed dispersal—their ecological benefits—is not what people have traditionally monetized or even recognized as a benefit to humans. Just as the U.S. Army used bees as weapons or as detectors of chemical and biological agents, the military also used domesticated homing pigeons to carry messages during wartime.[1] The military extensively used specially trained homing pigeons to carry messages in World Wars I and II but ceased using pigeons in 1957. But while G.I. Joe, a messenger pigeon with the U.S. Army Pigeon Service, won the Dickin Medal, Britain's highest award for animal valor, untrained and undomesticated wild birds win no such recognition for their ecological benefits.

Instead, the ecological recognition that most wild birds receive is that of being an "indicator" or "sentinel" species. Birds indicate environmental distress through their migration, nesting, and mating patterns, and their extinction or endangerment attests to these environmental issues. Birds serve as a particularly useful indicator species because of their incredible sensitivity to environmental changes, because they are well known, and because people have been documenting birds for so long.[2] Trish explained the general ways that birds help people understand our environment: "Birds tell us a lot about what the world is like, what our environment, the environment we need to live in is like—what our air, what our water, what our habitat is like."

Every birder I spoke to noted that birds are indicator species, and many of them used the idiom "canary in a coal mine" to describe this phenomenon. The phrase refers to the use of canaries to detect poisonous gases in coal mines. Miners used canaries rather than other portable animals such as mice, since canaries fly at such heights that they take in double doses of oxygen—one when they inhale, and one when they exhale.[3] Therefore, canaries are even more sensitive to airborne pathogens and provide an earlier warning than other animals that might be used to indicate poor air quality. The British government phased out the use of canaries in 1986, when it pioneered the use of an "electronic nose" to detect poisonous gases, but the image remains.[4] Birders noted that while no longer the literal canary in a coal mine, birds still indicate problems in many aspects of our environment. In our interview, Joy warned:

> There could be a time when we're back to almost what Rachel Carson predicted, a silent spring, because the birds can't accommodate the changes to the environment, and adjust, and evolve with the environment. It's not going to be a pretty picture. And with the climate change report Audubon released last fall, the prediction is somewhere around 300 species of birds will be either extinct or severely endangered within the next 50 years if we don't do something, because of the changing climate and their food systems moving and not being able to adjust. It's serious stuff. I think the trite saying is they're the canary in the coal mine, but they are the canary in the environmental issue of clean water, clean air, and food.

Joy cites birds as the primary indicator for the most dire environmental problem of our time—anthropogenic (human-caused) climate change. While birds as environmental indicators can tell us about the quality of water, air, and available food, all of these aspects are affected by climate change.

Birders see birds as environmental indicators, and specifically as indicators of climate change. As discussed in the previous chapter, birders' careful attention to birds and their data collection for citizen science projects help them see the effects of climate change on birds. Their yearly counts demonstrate that birds' ranges are moving farther and farther north, owing to climate change. Birders see these effects in their yearly counts as well as in their everyday walks—bird walks are often peppered with information about species birders see now, that they did not see in the past, because of the birds' increasingly northern ranges. Some birds overwinter farther north now because their food sources are increasingly available to them year-round, owing to a warming climate. Just as birds serve as indicator species, birds are also particularly useful for indicating climate change, as Rhonda plainly stated: "Birds are indicators of climate change, if you will, as much as anything else."

Since I conducted the majority of my fieldwork on birding walks in the northeast United States, I was able to observe birders discussing the northern expansion of birds' ranges. When birders discussed these expansions, they always noted that climate change was to blame. One walk leader referred to a Carolina Wren as "one of those climate change birds," since their range had expanded to create a 25 percent increase in New York State. And on several occasions in New York, Black Vultures confused birders who weren't expecting to see them, and whose field guides indicated their previous, more southerly range. Our walk leader, Bill, was confounded because the park's checklist didn't include Black Vultures, but he had seen them on several occasions in the park. Another birder said, "Well, all that is changing now, because of the climate," and Bill agreed.

Earlier in this book, I showed how birders might mistakenly identify a bird because of wishful thinking, which clouds their naturalist gaze. But sometimes, a mistaken identity can be due to field guides' range maps becoming outdated because of climate change. At a Project FeederWatch citizen science event at a nature center in late January, one birder said the previous weekend had brought about fifty visitors to the nature center. "One young man came up and said he saw a couple of Black Vultures in the tree. I said to myself, 'Young man, you're probably seeing eagles.'" The man then interrupted himself with a "NNNNT" sound, like a buzzer indicating a wrong answer on a game show. The man went on, "I went out to look, and sure enough, it was two Black Vultures." After telling us this story about his mistaken identification, he pulled out a Sibley field guide and pointed to the Black Vultures' range map. He explained, "Black Vultures are becoming much more common around here because of climate change. Their typical habitat is in the South, and you can see on this map there's just a tiny point up north, but they are becoming more and more common up here."

Birders also observed that they were seeing more and more species overwintering farther north, instead of flying south for the winter. On one walk, Diane, the leader, said, "This winter we had a lot of Catbirds overwinter in this park—this is a sign of the changing times." Another participant in Project FeederWatch noted that "with the warmer weather, we were getting a lot of birds we normally don't have in winter—Cowbirds, Red-winged Blackbirds.

Many of them just didn't leave and decided to winter here." These birders were discussing more than just a mild winter—they understood that climate change affected birds in less obvious ways and in ways that were not related to weather. For example, Luis, a walk leader, explained how global warming affected birds' food sources. "Breeding and food are no longer lining up," he said, "and this is true for all animals, not just birds. If a snake typically eats herbivores like rodents, who eat seeds, and the seeds aren't yet available when the snakes need to eat rodents, they die off."

Similarly, in our interview, Miranda explained how warmer winters don't encourage birds to come early, so if we have a warm winter and their food sources are already depleted, birds will be affected by this loss: "Climate change, it's really interesting to see how that's going to have an effect on birds, because people are so focused on weather, like, 'Oh, it's 70 degrees in February, are the birds going to come back early?' No, because they don't know what the weather is like here. They're in Central America. Or they're in South America. Nobody's telling them, 'Hey, you can go early.' But there are other things that can happen. Plants may bloom too early, the insects may hatch too early." Miranda explained that these elements all fit together and that they are each "a piece of the environmental puzzle."

Miranda's comments also clarify the differences among global warming, climate change, and weather. Global warming describes the recent, ever-increasing rises in the global average temperature, caused by greenhouse gases in the earth's atmosphere. Global warming causes climate patterns to change; this is climate change, which causes unpredictable weather patterns. Birders see these changes in the field, and they record data to demonstrate these changes on a larger scale, through their citizen science projects. Citizen science data informed the National Audubon Society's Climate Change Report, which Audubon president and CEO David Yarnold explained in a telephone town hall in January 2017. A caller asked the Audubon representatives for tips on bringing attention to the importance of climate change, and Yarnold explained that she could use particular birds that people care about to get them to better understand how climate changes affect those birds. He started by saying "it's really powerful to show people visual evidence about what's happening," and then he gave an example for the caller: "Go to the Birds and Climate Change Report on Audubon.org and look at the graphics that show what's happening to the habitat ranges where birds will be able to live. You can actually engage somebody in picking out a bird in your flyway, or in the state that you're in, you can get down to as particular as a species and look at what's happening to them. The graphics are really engaging, and one bird leads to another, and before you know it, you're absorbed in what those graphics are telling you and a whole lot less in what the politics say." Yarnold appealed to the pleasurable aspect of the naturalist gaze, and he encouraged the caller to take advantage of birds' charisma and the connections people have to their common birds to help people understand the effects of climate change.

Joy inherently understood this strategy as well, and in our interview she noted that people's love for birds actually makes it easier to talk about climate issues: "Birds are just a nice thing to get your head around, because we all like them so much. It's much easier to talk about birds than it is to talk about climate change. So if you can talk about birds, let's help the birds, then you're taking care of the other things too." Birds' charisma gets people to pay attention to birds, and birders who carefully observe birds through the naturalist gaze see birds as environmental indicators and indicators of climate change.

BIRDING AS A LIFESTYLE MOVEMENT

Conservation-minded birders turn their naturalist gaze back on themselves, as they work to lessen their own impact on the earth by making changes in their own lifestyles. Birders engaged in a variety of conservation practices in their everyday lifestyles, making this particular aspect of birding akin to participating in a lifestyle movement.[5] Lifestyle movements describe social movements that encourage people to make changes in their everyday lifestyles and consumption choices. Lifestyle movements target cultural codes and individual cultural practices by encouraging individuals to take individual-level actions. In contrast to traditional social movement participation, which encourages intermittent public engagement on an issue, such as participating in a protest, lifestyle movement participants engage in ongoing, private actions in their everyday lives. All of the birders I interviewed engaged in such actions. Many birders talked about keeping their carbon footprint low in a variety of ways, such as driving hybrid vehicles, recycling, purchasing carbon offsets, or insulating their house and windows. They also said they planted native

FIGURE 9. Window with decal to avoid bird strikes at Shin Pond, Maine Summer Retreat Program of the Humane Society of the United States (photograph by author).

plants in their yards or placed decals on their house windows to avoid bird strikes.

These birders didn't make these changes because they generally wanted to become environmentally friendly; all of the birders I interviewed made changes to their everyday lifestyle and consumption choices with the explicit goal of helping birds. Dawn listed a variety of practices she engages in, all of which could be considered general green-living practices, but she specified how they positively affect birds:

> Lots of ways to help birds! Simple ways that I incorporate into work a lot is to take action in your own sphere, whatever that is, a property that you own, a home that you're renting, a neighborhood park, your green space in the parking lot of where you work, whatever that is, we can really have an impact by planting native plants, reducing chemical use, simple things like recycling. You know, the boreal forest is a huge habitat ecosystem for northern species of birds, and a lot of our toilet paper and a lot of our paper, virgin paper comes from there. If we demanded recycled paper, we would have a real impact on that ecosystem.

> There's hard, complex ways to deal with these problems, and there's some simple solutions that people can take on their own, that they have complete control over. Like reducing the amount of mowed lawn and putting in habitat for birds to use. . . . Drinking shade-grown coffee is really critical, instead of sun coffee.

Dawn's comments exemplify several important issues for birds and birders. One is habitat, which she emphasizes can be improved through native plants and reducing mowed areas, and which I discuss in more detail in the next section. Another is her ability to tie together issues that may at first seem unrelated to birds. This ability to see connections between people and animals in a shared ecosystem makes up part of the naturalist gaze. Dawn understands that virgin paper products deplete the boreal forest, which is important for birds, just as shade-grown coffee helps keep the forest canopy, and thus birds' habitats, intact. Finally, Dawn emphasizes people's agency and ability to effect these changes, as she went on to say, "There's lots of ways to take action over your own space and your own lifestyle that can have a huge impact on birds in a positive way." These are changes Dawn believes people can and should take on behalf of birds.

Lifestyle movements also differ from traditional movements in that lifestyle movement adherents don't directly attempt to convert or persuade others to make similar changes. More often, they lead by example and explain why they make the choices they do. In these ways, lifestyle movements make typically private actions public, by explaining the reasoning behind them. Many of the birders I interviewed said they talked to others about their lifestyle in the hope of inspiring others to do the same, as Mona does: "Recycling, and using less energy. Planning my trips and not driving so much. Just some basic, easy stuff you can do. Talking to people. Just things that you do, once you 'get it,' and you live it, you want to share it with other people. And you have to do it in a nice way. You can't force somebody to do something they don't want to do, but you hope that they will see why you take the actions you take every day. And it's a positive thing."

Lifestyle movement adherents benefit from supportive social networks or groups of like-minded friends who encourage them to maintain their behaviors.[6] Birders who are members of Audubon chapters, like all of the birders I interviewed, benefit from support from their informal social networks as well as from the Audubon Society, which also encourages such lifestyle changes. Support from the Audubon Society helps birders who may not have such supportive social networks make environmentally friendly changes in their lifestyles. For example, Chloe lives in a remote rural area without recycling services, but she keeps her carbon footprint down in other ways. She said that living in a rural area helps her family live frugally, as opposed to "being in an actual urban area where you can always buy things."

> We're a ranch, we're just on the ranch, we don't go anywhere, we don't buy things, we don't drive anywhere. We do a lot of walking, we spend a lot of time outside, we don't have TV, so we use electricity minimally, we use water sparingly, talk to our daughters about it. We compost, we have our own chickens, we do beans, we have our own cattle, which is grass-raised. I guess those are all little things. One of my biggest challenges is recycling, because we don't have any recycling in Wyoming really to speak of. Major towns do, but where we are, we have to deal with our own trash. We have to take it to the dump personally, ourselves, so we use compactors, we compost, we have chickens and animals that we feed everything to, and we try not to buy stuff.

Binoculars help birders view closer images of birds, and the naturalist gaze helps birders see the environmental effects of their consumption practices. Birders reduce the physical distance between themselves and birds, and they reduce the metaphorical distance between themselves and the commodity chain.[7] Chloe exemplifies this as she went on to describe the closer relationship she has to her trash: "When you deal with your own trash, it's not going in a bag, and then it goes away. It's amazing how much you really pay attention to how much trash you generate." Recycling was a concern for Chloe, and since her area does not have a recycling service, she said that she will save her recyclables and bring them with her when she goes to town for staff meetings. Even in the absence of a community of neighbors or other close-by friends, Chloe maintains her lifestyle practices, since she knows other people are making the same changes for the same purposes.

Every birder I interviewed made some changes to his or her lifestyle, with a few beating themselves up for not doing more. One particularly impactful lifestyle change that surprisingly few birders mentioned as a current practice or a goal was reducing or eliminating meat from their diets. The National Audubon Society has published in its *Audubon* magazine about the environmental destruction caused by meat production.[8] Environmental scientists have shown that eating a plant-based diet represents a high-impact method for lowering one's carbon footprint.[9] In contrast, the same researchers showed easy and popular efforts such as recycling or using more energy-efficient light bulbs to be moderate- or low-impact practices for reducing one's carbon footprint. This same sort of conservation research informs birders' other environmental practices, but it did not seem to affect their dietary choices. Only two of the thirty birders I interviewed were vegetarian, and only one said she reduced meat in her diet.

The birder I interviewed who had reduced her meat consumption, Kay, said she found it ironic that as a birder she ate poultry: "I haven't eaten red meat since the 70s. I still eat chicken, which is ironic, considering I'm a birder. But I don't really eat a lot of meat. I eat only poultry when I do eat meat. But it's kind of the same thing. We have outsourced all the parts that contain responsibility." Some of my interviewees who did not engage in any meat reduction said they joke about this irony with other birders, as Ellie did: "And eating. We always make jokes about that. We're birders, we shouldn't be eating chicken. Or duck, for that matter."

The naturalist gaze does not compel meat reduction, much less vegetarianism. Nevertheless, the two birders I interviewed who largely abstained from meat were utterly confused as to why there were so few vegetarian birders. On the last day of the Audubon Convention, I interviewed Allen, who brought up vegetarianism without my prompting. He said, "It always amazes me that when you're with a group like this, how many people eat large amounts of meat." He went on to describe how his learning about the environmental impact of industrialized animal agriculture influenced his lifestyle changes: "When I started doing my thesis, I was reading about all these articles about the cattle industry, and the impact they have on the environment is just incredible. And how it's destroying bird habitat and all this wildlife habitat. I still eat meat maybe twice a month or so, something like that. But I almost sort of became a vegetarian, basically because of the environmental impacts. And the more I read about it, the more I think that was the right decision. I just felt like a hypocrite to be a biologist and then to be supporting an industry that was so destructive to wildlife." Allen said that he found himself to be hypocritical, being a wildlife biologist and at the same time supporting the meat industry, which was so destructive to wildlife. Speaking about this gathering of committed conservation-minded birders, he said he was "amazed" and "surprised" by the small number of vegetarians, especially among people "who are so concerned about the environment." He wondered if they knew about the effects of the meat industry on the environment, or "if they just never really put the connection together." To him, "it's something as obvious as an environmentalist or as a biologist that it's just something that to me just shouts out, 'this is not what we should be doing.'"

When I interviewed Marcus at the World Series of Birding, he likewise noted that he knew there wouldn't be any vegetarian food for him at the event, so he brought his own food. Marcus mentioned environmental reasons behind his vegetarianism, and he also compared birding to Buddhism: "Being a vegetarian, as you do it for years, you get more into it, and you start to have purpose behind it. Preservation of the environment. I went through many years as a Buddhist and the spirit of nonharming, which birding is totally nonharming."

These issues sometimes came up at Audubon chapter meetings, though indirectly. During a presentation about birding in South Africa, the presenter showed a photo of a Crested Guinea Fowl, and someone said "Dinner!" and everyone laughed. Later in the presentation, the presenter showed some photos of African Penguins and noted that they were endangered because their eggs were considered a delicacy, and thus their population was declining. The crowd then murmured disapproval at that fact. Why is eating endangered penguin eggs a travesty but the widespread environmental destruction caused by industrialized animal agriculture is not? Perhaps, like climate change, the scale and scope of industrialized animal agriculture are too large to fathom.[10] More likely, animal agriculture does not seem as directly related to birds as does shade-grown coffee. While a birder may immediately understand the relationship between shade-grown coffee and bird habitat, the relationship between industrialized animal agriculture and bird habitat is not as immediate. Meat production causes shifts in land use, converting forest to grazing lands, and such deforestation causes biodiversity and species loss.[11] Perhaps this extra step necessary for connecting meat production to wildlife conservation makes the relationship less clear. Thus, while birders can easily turn the naturalist gaze back on themselves and make lifestyle changes to benefit birds, not all such changes are as compelling or clear.

INSTITUTIONAL APPROACHES TO HABITAT LOSS

Birders also turn their naturalist gaze to institutions, as they seek institutional changes to help birds, and especially to improve bird habitats. Key efforts in

FIGURE 10. Invasive species sign at Assabet River National Wildlife Refuge in Massachusetts (photograph by author).

this area include planting native plants for birds and removing invasive plant species. As discussed in chapter 1, birders receive education about various plants or habitat features when walk leaders explain why the group will see a particular species of bird in this particular habitat. As birders develop their naturalist gaze, they learn where—in which habitats—to expect to see specific birds. But birders were not only interested in habitat for reasons of finding and identifying birds—they know that habitat loss poses a threat to all birds. Birders seek to improve bird habitat in their own backyards, in their local areas, and at the state and national levels.

Birders recognize invasive plants as a threat to birds' habitat. Environmental sociologists have tied having a higher number of invasive species to having more economic development, and at the same time, invasive plants are costly: biologists estimate that invasive species cost the United States $120 billion per year.[12] While invasive species, especially plants, come to an area in a variety of ways that birders can't necessarily combat, birders attempt to make positive changes in birds' habitat where they can, by removing and eradicating invasive plants.

The naturalist gaze is integrative, viewing flora and fauna as interconnected elements in a shared ecosystem. The naturalist gaze is also evaluative, judging flora and fauna in that ecosystem, as well as instructive, meaning walk leaders teach walk participants to view the natural world through the naturalist gaze. Birders judge plants in a manner similar to how they judge birds, as described in chapter 5. But, unlike birds, plants don't have agency. Bird walk leaders often point out invasive plant species to participants, teaching them about the dangers of such plants for native birds. On one walk, Arnie brought along a plant specialist to help identify the plants that the group saw on its walk. The plant specialist simply identified the plants, and Arnie provided the commentary. Thus, in addition to learning about the different trees and native shrubs, birders also learned about the dangers of invasive plant species. Arnie said, "We don't need to waste our money planting new trees, we just need to get rid of the invasive species that prevent our native trees from growing."

Sometimes, walk participants enjoyed the invasive species of plants when they provided tasty fruit for birders to sample. When walking on a path through the forest on a bird walk, Garrett, the walk leader, pointed out several bushes of wineberries. "These are edible—who wants to try one?" The group stopped and Garrett picked some of the ripe, dark red berries to share. An Asian species of raspberry, they resemble North American raspberries but are invasive, nonnative plant species in North America. As the group sampled them, Garrett mused, "We're doing the land a favor by eating the berries, because wineberries are invasive, so we won't be spreading the seeds like the birds would. If we ingest them, they won't be going back out into the world to plant new seeds."

Birders not only point out invasive species while on birding walks. They also engage in advocacy with local park commissioners to encourage them to eradicate invasive plant species in order to provide good habitat for local nesting birds. In Amy's case, her knowledge of botany and her fight against invasive plant species led to her getting involved in her local Audubon chapter. Amy earned a master's degree in English literature and is a former English teacher. She retired from teaching to take care of her children. During this time, Amy started taking botany classes. Speaking of her local park, she said, "There's a lot of invasive plants, but there are patches of Big Bluestem, and Indian Grass, and Switchgrasses, and if you sort of squint a certain way, you can almost think that you're in the middle of the tallgrass prairies." When Amy noticed the local park mowing the grass, this worried her as someone who wanted to keep the native prairie grasses intact. Amy started talking to the birders who also frequented this park, and she learned that grassland birds nested on the ground. This doubled her concern, as the mowing would damage the native grasses and kill the birds that used them for nesting areas. Amy described for me what happened next:

> I was really upset about that, so I wrote a letter to the county parks commissioner, and I said you advertise [this park] as a grassland bird habitat, and you're destroying it. You're mowing these grasses. I found an example of a landfill in the Midwest, where they had stopped mowing and had this resurgence of all of these grassland birds. So, I put all of this information, I researched it, sent it to the county parks commissioner, copied Diane, and that's how I got involved. She said come to a meeting, there was a real reaction in the parks management to my letter. And from that, we started this group working with the county to try to improve the habitat at [this park]. That's how I got involved.

Amy's interest in birding grew out of her interest in native plants, and she became involved with Diane and her local Audubon chapter because of her advocacy for the native plants at this county park. She said, "I thought I might as well try to learn some birds, because it will help me appreciate what the connection is between the plants and the birds." Amy had developed a naturalist gaze through her botany courses, so she was able to understand the relationship between the grasslands and the birds that nested there, as she said that becoming a birder, in addition to her botany knowledge, "really has improved my understanding of all of the connections in the ecosystem out there." Her participation in the local Audubon chapter helped further this understanding and, she said, "I think it's made me a more effective advocate, because I really have a much better idea of what the birds need."

I witnessed this local Audubon chapter's ongoing relationship with the county parks commission on a walk with Diane, in the same park Amy had described. As we started walking, we passed a young man on a truck who had a bunch of mowing gear with him. He was unloading a string trimmer, and Diane took notice. She asked the group to wait a moment, and she went to talk to this worker. I couldn't hear what she was saying, other than "Audubon," but I saw her get out her phone. At first, I thought she was getting permission to walk through this area of the park, since it didn't have a footpath. But when she came back, she explained what was going on: "They aren't supposed to be mowing in the area, because it's nesting season. There are two different county offices that control the park, and it's hard to get them both to understand and work together. The two heads [of the offices] are talking more than they ever had before, but there are still some kinks to be worked out." She explained that they had to keep the paths clear up to the white poles that were sticking out of the ground, but that the county had agreed not to mow during nesting season. She said, "He's probably just going to wait for us to leave, and then go ahead and mow." She then stopped and asked us to wait one more minute—she said she wanted to take care of this. She sent a text, and then she told us that she texted the commissioner and that she'd call him later. She said, "Let's go on with the fun stuff, I don't want to keep you all here for this." Throughout the rest of the walk, she worried about this out loud from time to time, saying, "They said they weren't going to mow," and "I really hope they don't mow." (When we returned, they had not mowed.)

Diane and Amy use their naturalist gaze to understand bird space and bird time, as previously discussed in chapter 2. They understand that grassland birds nest on the ground, and that since it was nesting season, the parks commission should not be mowing at this time—they could destroy nests and kill nestlings. This danger was exacerbated by the fact that this was the only grassland habitat around for many miles. The birders appreciated the uniqueness of the habitat, but the county parks commissioners did not seem to view the land through the same naturalist gaze. Diane, Amy, and other birders in their local Audubon chapter therefore worked with these county parks commissioners to protect the grassland habitat for these nesting birds.

Birders know that habitat loss represents the number one threat to birds' survival, and birders work to improve bird habitat by turning their naturalist gaze to institutions affecting habitats at the local, state, and national levels.[13] Habitat fragmentation describes the breaking up of birds' habitats along their migratory routes, and this fragmentation also threatens birds' survival. Most of the land in the United States (72 percent) is privately owned. The U.S. government owns about 640 million acres, or about 28 percent of the 2.27 billion acres of land in the United States.[14] While most of the government-owned land is managed for recreation, preservation, or the development of natural resources, the majority of privately owned land is used in ways that break up or eliminate the natural habitats used by birds. Privately owned land is developed and used for housing or other buildings, logged for timber, or converted from grassland to agricultural fields.[15] This leaves resident birds with few places to feed, nest, and breed, and habitat fragmentation leaves migratory birds with even fewer places to stop along their route.

Birds don't migrate for the same reasons that wealthy people "winter in Florida" or "summer in Maine." They aren't trying to avoid extremely cold or hot temperatures; they're trying to find food. Birds migrate if their species relies on food that is available only at certain times of the year. If birds eat insects, nectar, or seeds, they often need to migrate in search of food. Birds' migration follows regional flyways, or paths between the Northern and Southern Hemispheres. In the United States, birds follow one of four flyways from the Caribbean or Central or South America to North America: Atlantic Flyway, Mississippi Flyway, Central Flyway, and Pacific Flyway. In Eurasia, Africa, and Australasia, birds may take the West Pacific Flyway, the East Asian–Australasian Flyway, the Central Asian Flyway, or one of three African-Eurasian Flyways: the East Atlantic Flyway, the Black Sea–Mediterranean Flyway, or the West Asian–East African Flyway. These flyways can span entire continents and cross oceans, and thus migratory birds—about 40 percent of all species of birds—face even more challenges if the habitats they rely on during their migratory period are fragmented or destroyed. To be clear, the resident birds that make up about 60 percent of all bird species also face significant consequences from habitat loss; the problem is simply increased for migratory birds that cross thousands and thousands of miles of habitat during their life cycle.

Over half of the world's migratory bird populations are declining, and researchers have found that 91 percent of migratory birds have inadequate habitat across all stages of their annual cycle (resident, breeding, nonbreeding, passage).[16] Resident, or nonmigratory, birds fare somewhat better, with 55 percent of them having inadequate habitats.[17] More specifically, researchers have found that birds' wintering grounds face the largest habitat loss.[18] Currently, the wintering grounds habitat loss is due to development, or people converting forests to grasslands for grazing or croplands for agriculture, but scientists predict that by the end of the twenty-first century these habitat losses will be worsened by the long-term effects of climate change. They project a decline in rain of 20 percent or more on their nonbreeding grounds, and greater warming on their northern breeding grounds and during autumn migrations, meaning further reduction of available habitat and food.[19]

Climate change will especially affect shorebirds' habitats, as it causes sea levels to rise. One study of ten shorebird taxons that primarily migrate via coastal habitats on the East Asian–Australasian Flyway found that sea-level rise will inundate 23–40 percent of these birds' intertidal habitats along their migration routes.[20] These reductions in stopover habitats will concomitantly create a 72 percent reduction in the population of birds that migrate through those areas.[21]

Birders recognize the links between habitat loss and extinction. On one bird walk, the group members began discussing the Ivory-billed Woodpecker, since they had just hosted a talk about the (presumed) extinct bird at their previous meeting. One man asked, "Can its extinction really be blamed on people?" and Tom replied, "Yes, it's from loss of habitat." The other man went on to compare its extinction to previously extinct birds, and how people didn't realize that extinction was even a possibility: "People didn't think they could wipe out the Passenger Pigeon, since there were so many of them, and yet we did wipe them out, entirely." Tom agreed, and explained: "Back when people first started calling themselves conservationists, back in the 1700s, what they meant was they would shoot birds and then identify them. It's like when we go out for the Christmas Bird Count, but instead of counting birds, they'd see how many they could shoot." He went on, "People saw so many birds, or so many trees, and they never thought they could disappear." The concept of extinction exemplifies culture lag—it was happening before people could comprehend it, much less put a term to it. One day they were shooting Passenger Pigeons by the hundreds, and the next day they were all gone.

In our interview, Nancy made this link as well, and she tied it back to habitat loss: "With the Passenger Pigeon, when they existed, one in four birds, when you looked outside, was a Passenger Pigeon. Now there's none. And if we continue down this path of destroying the environment, you're going to have less and less birds, which are going to reflect the destruction of the environment. The Passenger Pigeon went away because of humans, but other birds are declining, again because of humans, but in a different way. We're getting rid of their habitat. That's the biggest problem right now, is the habitat loss." These structural issues, along with the small amount of land protected by the federal government for conservation, led birders to argue that the United States needs to make conservation a priority at the national level. They turned their naturalist gaze to institutions like the federal government, as Amy did. She said, "We need someone top down on the national level, like another Roosevelt, to make conservation a priority." She gave the example of planting trees and said in the 1930s, the federal government planted thousands upon thousands of trees, but "for some reason it's a huge, big deal for the parks department to put trees in. Well, they did it during the Depression. Why can't you people do that now?" Amy said "they just don't have the will," since no one at the national level is pressing for it. She continued, "I don't think Hillary Clinton's going to be a Roosevelt, but we need someone like that." Our interview clearly occurred before the 2016 presidential election in the United States, but Amy's point still stands, that the United States hasn't had a president who made conservation a real priority since Roosevelt, and the country would benefit from more support from the federal government to make these significant changes.

Birders also recognized the need to implement conservation programs at the state level. Whether on their own or as part of a local chapter, birders told me about their work contacting state legislators or working with state parks to protect birds. Many of the state-level programs of National Audubon work with governors and state legislatures on a variety of issues, and they manage Audubon's Important Bird Areas. Joy explained the importance of these areas in her state:

> There are about ninety-five Important Bird Areas in North Carolina that are locations where birds need that space at some point in their life cycle. It might be for breeding, it might be for winter grounds, it might be their year-round residents. But those are some of the areas that we're losing because of habitat loss. We can do a lot to stop some of those things, and the more we do to protect against climate change and save land, the more we will help birds. And that's important in eastern North Carolina right now, because, I don't know if you know, they're foresting and cutting trees all across eastern North Carolina for the wood pellet industry. We're shipping them to Europe. We're losing thousands of acres of trees in eastern North Carolina.

Joy emphasized the need to protect these Important Bird Areas because of habitat loss, citing climate change and forestry as dangers to birds' habitats.

Finally, birders worked to protect habitat at the local level, including in their own backyards. Birders turned their naturalist gaze to dead trees in people's yards, seeing them as habitat, whereas nonbirders might see them in other ways, as Kay explained: "A dead tree becomes this liability instead of a habitat." Homeowners may see dead trees as problems to be solved, but birders see them as habitats that birds need. Rhonda chalks this up to cosmetic reasons: "If you cut down all the old trees that look so yucky, because they're trashing your pretty property, there's no spaces for certain birds to nest anymore, or for woodpeckers to get grubs anymore, that grow in these old trees. So you will need to be more aware of the impact of their actions. And their cosmetic decisions." Other birders talked about helping their neighbors learn how to plant native plants in their yards, as a way of attracting birds and butterflies.

As noted earlier, most land in the United States is used for agriculture (crops, grazing, and pasture) or logging. In fact, only 6 percent of U.S. land is developed.[22] However, the ecological impact of development is disproportionately high in comparison with those other types of land use. Further, since four out of every five people in the United States live in an urban area, these ecological impacts intensify.[23] This is why the Audubon Society and birders encourage people to plant native plants, make their buildings friendly to birds, and build structures to help birds, such as nest boxes: these actions make a significant difference in urbanized areas. By turning their naturalist gaze to local, state, and federal institutions, birders work to protect bird habitats in a variety of ways.

CHILDREN'S ENVIRONMENTAL EDUCATION

After turning the naturalist gaze on themselves and on institutions, birders try to develop the naturalist gaze in others. Specifically, birders try to develop the naturalist gaze in children so that they will "create more conservationists," as they put it. Bird walks provide informal education, during which time birders learn how to bird and develop their naturalist gaze. Most bird walks are aimed at adults, though some cater to children and young adults. Audubon centers and chapters hold formal education programs for adults, such as workshops on how to bird by ear. But most often, Audubon centers cater to children and develop many formal educational programs for children.

Several of the birders I interviewed worked as educational outreach coordinators for their local Audubon centers, and they taught birding programs to both adults and children. Cynthia described the subtle approach she takes to teaching children about environmental issues: "I don't really define it as environmental issues when I'm talking to children; we're just kind of outside exploring and learning together. When I do programs with neighborhood associations or adults, we talk about things like sustainability and some of the environmental issues that they're having." While Cynthia may not define her talks to children as being about "environmental issues," they still learned about the environment in these programs. Environmental educators don't shy away from teaching children about the tough conservation lessons of the past, and many of them taught children about extinction. Audubon centers host regional or national child-oriented programs, such as various summer camps hosted by Audubon centers across the United States, or the Audubon Adventures program, which provides curricular programming about birds and their environments for teachers to use in their classrooms for grades 3–5.[24]

Birders don't stop teaching children about birds once they leave their Audubon center or their classrooms. Many birders I spoke to said that they were always teaching children, in informal ways, about birds and environmental issues, as Chloe did. She said that when she takes her two young daughters out, she always points out birds: "What are they, do you see them, do you hear them, what do you think about them? I like to know what everything is, but I'm also not a person that's going to take away that discovery for someone else. Do you see that bird? And then if they ask questions you can fill them in. Otherwise, I don't think it's so important to be like, 'Well, that's a brown-chested chat' or whatever [*pauses*]. That's not an actual bird." Chloe takes her children on nature walks and uses that as an opportunity to teach them about birds and, more importantly, develop their own curiosity. Chloe's walks are different from nature walks for adults, where much of the time is spent calling out bird identifications. Chloe said she doesn't want to "take away that discovery for someone else." She wants her daughters to see birds, ask questions, and develop their own interests.

Birders understand the importance of these kinds of connections for getting people, especially children, to care about the environment, as Dawn said: "Trying to increase the amount of people who care about birds and want to protect them, and having the bird either sighting up close or really unique or to actually hold the bird in your hand for bird banding or something like that is a life changer. And a life changer in regards to potentially being a conservationist, and that's what drives me, is trying to create more conservationists, or increase, or enhance conservation action." Dawn said that she tries to "create more conservationists" through their interactions with the natural world.

Birders also understood that this kind of hands-on experience was especially important for children, since abstract, far-off animals aren't the best way of getting them interested, as Kay told me. She said that birding could be an entry point for children, to show them "there is a wild world depending on us to preserve it." She continued: "That could be the entry point. And then maybe you care about the otters and the polar bears. Because what does a New Jersey kid know about a polar bear's habitat diminishing? Like the kids in Alaska that my brother deals with know all about it. But the kids here, why would they even care if there's never a polar bear again? There's no more white rhinos. That doesn't affect me. I don't care. But some day it's going to." Teaching children in New Jersey about polar bears isn't as effective as teaching them about a bird that they can see themselves, in their own backyards. Birders don't only emphasize common birds because they represent the most populous species on any given walk. Rather, birders encourage others to appreciate common birds since they help people connect to their immediate environment, as Kay said.

Andy said that, as a former educator, he believed children intrinsically love nature, and that teaching them about nature at a young age will create a lifelong appreciation for nature: "Kids are innately kind of fascinated by cool stuff in nature. If you show it to them, they'll gain an appreciation for it. And if they don't get disconnected from that, if there's some kind of maintenance, if they maintain their connection with the environment through educators or a teacher, then they'll learn. They'll be the protectors in the future, and be the advocates for preserving land, or conserving land, that kind of thing." Andy and several other birders believed that getting children interested in birding would help "create conservationists" in the future. Andy's approach to showing children the "cool stuff" in nature also highlights a strategy for these environmental educators: keep it positive. Even though these birders talked about extinction with children, they didn't present it in a manner intended to shock. Just as zoos hide the less family-friendly aspects of animal life, such as feeding carnivores or animals' deaths, Audubon educators avoid talking about the "doom and gloom" of environmental issues when teaching children about birds and the environment.[25] Instead, they focus on what people can do to help the environment, as Nancy explained, when she talked about an exhibit they were planning at their Audubon center: "The museum's not going to be doom and gloom. A lot of it will be the success stories that have occurred because people have realized there was a problem, stepped in, and here's what you can actually do. This is what's been done, and we have to be aware of what's going on to stop things before they become out of control." While education researchers have found that learning about death is an important way for children to develop a relationship to the natural world, these birders found a way to present the realities of extinction to children while emphasizing how they can help birds and the environment.[26]

ADVOCACY

The naturalist gaze moves birders from looking at birds to taking action on behalf of birds. Once birders see, through the naturalist gaze, the connections between people and animals, they work to protect the natural world. The naturalist gaze culminates in birders engaging in advocacy for environmental and wildlife conservation. Many people hold a stereotype of birders as quiet and solely interested in observation, but this view misses the entire history of the Audubon Society as an advocacy organization founded to protect birds. The conservation-minded birders I interviewed carry on this legacy and engage in advocacy at the interpersonal, local, state, and national levels.

At the interpersonal level, many birders found that low-pressure, one-on-one conversations help educate people about environmental issues that they weren't previously interested in. Trish said she finds it easy to talk about the environment with other birders—they "naturally talk about that," she said. Trish said she has to creatively come up with ways to bring it up in conversations with her quilting group: "With my quilting friends, my themes tend to be the natural world that I weave into these things." Interspersed between conversations about where to buy thread, Trish said she will bring up "what's happening to our local park" or "school buses being hybrids." Similarly, Miranda said she tries to bring up environmental issues to others, but in a delicate way: "I'm constantly trying to get people involved and point things out to them, and not in a pushy way, but just, 'Oh, this is interesting.'" No matter the group, the birders I interviewed said they try to bring up wildlife and other conservation issues with as many people as possible.[27]

Other birders I interviewed engaged in advocacy in their local communities. Diane talked about her involvement with a local town that was trying to deal with coyotes: "They wanted to make a coyote management plan, which is absurd, even by its title—it's goofy for one town to do that, because it's obviously a statewide, regional thing. So, I spent a lot of time, and other people in Audubon spent a lot of time talking, going to meetings, writing letters." Viewing coyote management through the naturalist gaze, Diane saw that efforts to deal with coyotes on an individual town basis would never work, since nature doesn't adhere to town borders. Instead, Diane pushed for this to become a regional effort, and she said the result was eventually successful. Earlier in this chapter, I discussed Amy's work with her local and county parks on native and invasive plant issues, in which she viewed grasslands through the naturalist gaze and saw them as something to protect for breeding birds, and not something to mow simply because it was summer. Miranda, a leader with her local Audubon chapter, said she spread her efforts beyond Audubon to work with other environmental nonprofits, since she saw their work as complementary:

> I'm involved with a couple of local nonprofits in my town that advocate for smart growth practices and use of development. Another group is focused on the connection between open space and trails and the local economy, and how the two kinds of things are not at odds with each other, and really benefit and support each other. So that's lots of going to town board meetings, and submitting comments on different proposals, and educating the public, or educating business owners, or elected officials about different environmental practices. And then clearly at work, we are encouraging our members to advocate for a county budget that supports county parks and curators that are focused on our natural resources on county properties, getting a state budget that's passed that has good support for the environmental protection fund.

Miranda worked in tandem with these other local environmental groups at the county and state levels, and she said she contacts her elected officials at the federal level to encourage them to "vote for the environment." Miranda worried about "rollbacks" to "the 1950s" and said, "It's a shame, because if these businesspeople really understood that a clean and healthy environment is good for business, but they don't necessarily look that far into the future. They're looking at their short-term gains and not the long-term gains."

Miranda saw the connections among local, state, and federal environmental policies, and she worked at each level to protect the environment for people and for birds.

The National Audubon Society leads the charge for environmental and wildlife conservation at the national and international levels, and it works with its state and local chapters to link advocacy efforts at the state and local levels. National Audubon engages in management of working lands, partnering with farmers and state and federal governments. The Birds and Climate Change Report includes evidence of the effects of climate change on birds and strategies for lobbying elected officials and creating and maintaining Important Bird Areas throughout the United States. Audubon's "bird-friendly communities" initiative encourages individuals and communities to plant native plants, build bird-friendly buildings, and build nest boxes for birds. Audubon works to protect and restore coastlines, as well as engage in water management and water quality monitoring. Its citizen science projects count and monitor bird populations, and it uses these data in its advocacy and lobbying campaigns. Audubon Centers throughout the United States, and especially in urban areas often underserved by environmental groups, provide education about bird and climate issues. Their nature camps and Audubon Adventures camps teach children at a young age to appreciate and protect birds and the environment. The National Audubon Society in the United States partners with other countries in their flyways, in Central and South America and the Caribbean, on conservation projects, and on ecotourism. These partnerships attest to the borderless quality of bird conservation, akin to the early twentieth-century efforts of Canadian and U.S. duck hunters to preserve the Prairie Pothole Region, seeing the Midwest of both countries as a shared ecological commons.[28]

These material and structural solutions, implemented at various levels, are led by the National Audubon Society and by state and regional Audubon chapters, and they are implemented by conservation-minded birders like the ones I interviewed and observed. For example, in a national town hall telephone meeting with the National Audubon Society, Chief Network Officer David Ringer encouraged participants to do "absolutely anything" to spread the word about protecting birds. He then laid out several examples of what would be considered mainstream tactics in an activist's repertoire: "Look in the newspaper for stories about things you care about—your newspaper might run a story about Senate confirmation hearings for the new EPA administrator. Write a letter to the editor, just 200 words, and say, 'I'm a resident, I read this story, here's what I'd like to tell my fellow residents.' That's a great way to speak out publicly. Also, talking with family and friends, posting on social media—find an article on the Audubon website and post it to Facebook. And, get involved in your local Audubon chapter—many of them run speaking programs and educational programs, other ways to get active and reach people."[29]

Birders, because of their own "tastes in tactics" and because of the tactics that National Audubon encourages, engage in mainstream, institutionalized forms of advocacy.[30] In our interview, Allen described how he engages in a variety of these mainstream tactics on behalf of birds. He said he writes letters, sends emails, and makes phone calls to his legislators and to heads of agencies working on particular environmental issues. As an Audubon chapter leader, he said he also passes the information on to the chapter listserv. He also makes his voice heard at the local and state levels: "I've spoken two or three times at county commissioner, town council meetings, probably not as often as I should, but I've done a little bit of that, too. And then I volunteer with the land trust. I've been to lobby day, where all the land trusts in the state have some of their people, staff or volunteers, go to Raleigh and talk to the local, the state, and their state senator and state representative."

Allen listed a variety of tactics, such as signing petitions, lobbying, writing letters to the editor, and speaking at local meetings. National Audubon encourages these tactics in its "action alerts" that members can receive via email. In these action alerts, members learn about an issue and can simply sign a form letter to send to their elected representatives, or they can edit the message to make it their own, as Mona described. She said she created her own template using her own words instead of what was in the action alert, so that "when these things come out, you can tweak it just a little bit to make it your own, your own thoughts." Rather than just clicking a button to pass it on, she said, "it's much better if you do it in your own way." Whether or not birders personalize Audubon's message, the nudge from national or state Audubon programs to take action was important for many of the birders I interviewed, like Rhonda, who said she participates "only at the request of New York or National Audubon. I don't initiate, and on a personal basis. I participate if somebody else initiates it."

Finally, my interviewees made it clear that birding made them advocates—they developed their naturalist gaze and then sought to improve the natural world for birds and other living beings. Joy said, "Birding has made me appreciate nature more than ever, and it has really made me become an advocate for nature." Her appreciation turned into action for birds. She listed several of her advocacy endeavors: "I moved into writing lots of letters, calling legislators, on Monday, for the first time, I lobbied on Capitol Hill for the North Carolina representatives and senators, so it has made me become politically aware of what's going on." At the state and local levels, she visits her state legislators and is involved in the city council to encourage members to "adopt a policy of turning out lights on city-owned buildings, to avoid bird strikes." In each of these cases, the love of birds and birding comes first, and it sparks birders' environmental and wildlife conservation advocacy.

The naturalist gaze moves birders to take action for conservation efforts, and this chapter has demonstrated the material effects of the naturalist gaze. As birders develop their naturalist gaze, their careful attention reveals birds as environmental indicators and indicators of climate change. Birders turn their naturalist gaze back on themselves, making changes in their own lifestyles to become more sustainable. They turn their naturalist gaze on institutions, seeking to improve habitats for birds. They also encourage children to develop a love of birding, to "create more conservationists." Finally, birders engage in advocacy efforts to pass and enforce laws protecting nature and birds. In the conclusion, I revisit the theoretical and practical arc of the naturalist gaze.

CONCLUSION

WHAT WOULD JOHN JAMES AUDUBON think of these contemporary birders? What would he make of these people who organize walks, count bird species, and engage in bird conservation, in his name? He certainly would appreciate the advances in bird photography, to be able to capture the beauty of birds in even more lifelike ways than his amateur taxidermy for his bird paintings. He also would likely appreciate the ease with which birders can travel, to see more birds, especially birds in different flyways. Modern travel would certainly provide a welcome change from his dangerous voyage to Labrador, which he described in his journals: "We were drawn by the current nearly upon the rocks; but the wind rose at last, and we cleared for sea. At three o'clock it became suddenly so foggy that we could not see the bowsprit. The night was spent in direful apprehensions of ill luck; at midnight a smart squall decided in our favor, and when the day broke on the morning of June 8, the wind was from the northeast, blowing fresh, and we were dancing on the waters, all shockingly sea-sick."[1] Not to worry, during the voyage Audubon still "landed for a few minutes, and shot a Hermit Thrush."[2]

More importantly, what would Audubon think of how human society has wreaked the ecological disaster of anthropogenic climate change, which affects birds' ranges, habitat, food sources, and sheer existence? Even though naturalists of his time did not have an understanding of the human-made causes of extinction or of climate change at all, I think he and the other naturalists of his era would be disappointed in us as a human species. We have irreparably damaged the natural world they loved. At the same time, I also think he would take inspiration from these modern birders and their naturalist gaze.

THE NATURALIST GAZE

In this book, I have shown how birders develop the naturalist gaze, which comes from watching wildlife with a consideration of what is in the best interest of native wild animals in their natural habitat. Birders first develop a naturalist gaze through bird walks, which highlights the *pleasurable* aspect of the naturalist gaze—it's fun to watch birds! People go on bird walks because they want to develop a new hobby and learn more about birds. The naturalist gaze is also *instructive*, as walk leaders use it to teach new birders how to watch wildlife. Birders learn new aspects of the environment to pay attention to, contrary to all of the lessons that culture teaches us about ignoring "unimportant" background elements.[3] Birders learn to pay attention to the slightest of movements and the faintest of sounds, and they become immersed in the practice of paying close attention. At the same time, new birders also learn, through the naturalist gaze, that their new hobby is much more scientific than it first seems. The naturalist gaze is *informed* by field guides, by scientific research, and by environmental and wildlife conservation information.

The naturalist gaze affects what birders see. Birders pay attention to all the elements of the environment, and they put those components into categories that other people might not even know exist.[4] To those ignorant of botany or ornithology, invasive phragmite is just a "pretty plant" and nonnative Mute Swans are just "majestic birds." Birders view them through the naturalist gaze and see them as potentially detrimental and dangerous to both the natural environment and birds. This judgment comes from the *evaluative* aspect of the naturalist gaze, assessing the flora and fauna in an ecosystem on the basis of scientific information. Birders filter what they see through what they know, and they appraise aspects of nature for their environmental appropriateness and not simply for their physical beauty. This judgment also occurs because the naturalist gaze is *concerned* with the health and well-being of wildlife in its natural habitat.

The naturalist gaze also affects what birders do. Birders don't just sit on this information they've amassed; they use it to help protect birds, other wildlife, and the environment, for people and for birds. This action orientation is spurred by the *integrative* element of the naturalist gaze, which encourages birders to view people and animals as part of a shared ecosystem. Birders use their careful attention to detail and knowledge of species to conduct citizen science projects, which help conservation scientists study the effects of anthropogenic climate change on birds. Birders also use their knowledge of what helps and what harms ecosystems to help birds. Birders engage in direct actions to ameliorate the environment, such as planting native plants and removing invasive ones. Birders also change their own lives, engaging in personal actions to make their own lifestyles more sustainable. Finally, birders engage in environmental advocacy, lobbying legislators and working with national, state, and local governments to pass laws protecting birds.

LOOKING AHEAD

In this final section, I issue some challenges to readers, including sociologists and birders. The year 2018 marks the "year of the bird," in honor of the centennial of the Migratory Bird Treaty Act (MBTA).[5] This wildlife protection act was signed into law in 1918, on the heels of the Audubon Society's early conservation efforts to halt the use of birds and bird feathers in hats, described in the introduction to this book. The hunting of birds for their feathers and for sport drove many species of birds to the brink of extinction, including the Snowy Egret, whose numbers bounced back after the passage of the MBTA. The Snowy Egret now serves as the symbol of the National Audubon Society.

To celebrate the year of the bird, the National Audubon Society, BirdLife International, the Cornell Lab of Ornithology, National Geographic, and more than one hundred other organizations are joining together to publish new books, magazine articles, and television shows celebrating birds. They're creating new digital tools and migration maps to follow birds. They're offering new international birding expeditions, local educational events, and museum exhibits to teach people about birds.

At the same time that birders and other conservationists are celebrating birds and the centennial of the MBTA, the current administration is actively seeking to dismantle it and other protections for wildlife and the environment. On April 11, 2018, the Department of the Interior issued a new interpretation of the MBTA to wildlife law enforcement.[6] This new interpretation ends the ability to penalize corporations that "inadvertently" kill birds through routine practices such as oil waste pits or catastrophic accidents such as the BP *Deepwater Horizon* oil spill, which killed more than one million birds.

The list of challenges to environmental protections from the current administration looms long, including lifting bans on trophy hunting, dismantling the Endangered Species Act, defunding the Environmental Protection Agency, loosening regulations on toxic air pollution, and removing any references to climate change in the Federal Emergency Management Agency's strategic plans.[7] Can birders and other conservationists encourage legislators and cabinet members to view the environment through the naturalist gaze instead of through a capitalist gaze, where they make decisions on the basis of what is best for the stock market? Now, perhaps more than ever before, it is crucial for those who care about the environment to work on behalf of wildlife and environmental conservation.

Anthropogenic climate change provides a serious threat to birds' and people's existence. Birds are the proverbial canary in a coal mine, and we need to pay attention to what birds tell us. Scientists from the National Audubon Society and the National Parks Service have published a new study on birds in national parks.[8] They predict that by the year 2050, climate change will affect the breeding and feeding ranges of a quarter of all bird species in U.S. national parks. They project that Bald Eagles will no longer visit the Grand Canyon in winter, nor will the Common Loon breed in Acadia National Park. Those birds will instead be found farther north, where the climate suits their needs.

The conservation-minded birders I studied for this book work on the front lines of studying birds and providing data to conservation scientists through their citizen science projects. These birders work for environmental conservation in other ways: in addition to working with local, state, and federal governments, they have taken steps to make their own lifestyles more environmentally friendly. However, most of those steps have, scientifically speaking, only a moderate to low impact in reducing one's carbon footprint.[9] I challenge birders and other conservationists to take a further step and embark on a high-impact method for lowering their carbon footprint by reducing or eliminating meat from their diet. The National Audubon Society promoted "the low-carbon diet" in its *Audubon* magazine in 2009, shortly after the Food and Agriculture Organization of the United Nations released a study showing that industrialized animal agriculture contributes 18 percent of all greenhouse gas emissions, more than the 13 percent produced by all transportation, combined.[10] It is vital to target government and corporations through lobbying, boycotts, and petitions, as well as other methods of protest and resistance, both normal and disruptive. Corporations' and governments' contributions to damaging the environment certainly outweigh our own personal ones. But we can still engage in traditional forms of activism while at the same time taking steps to reduce our own personal contributions to anthropogenic climate change.

What else do I hope readers have gotten out of this book? Primarily, I hope that birders have recognized a bit of themselves in what I wrote. I also hope that I did justice to your important work. Keep working to protect birds, and keep sharing your love of birds with others. I hope that sociologists will see human-animal-environment relations in a new way. The environmental sociologists William Catton and Riley Dunlap exhorted sociologists to usher in the New Ecological Paradigm back in 1978, but sociologists, and society as a whole, still tend to neglect the animals that share this world with us.[11] In this book I've shown how birders have embraced the perspective of the New Ecological Paradigm, as they view people and animals as part of a shared ecosystem. Finally, I hope that all readers will be inspired to get out and enjoy nature. The natural world holds many wonders, even in our own backyards. We just have to pay attention to it.

ACKNOWLEDGMENTS

This book would not have been possible without the generous participation of birders. Their dedicated efforts to protect birds are crucial for the entire natural world, and it's been an honor to document their work. The birders in the local Audubon chapters with whom I went birding for the past three and a half years shared their work with me, expanded my perspective on the natural world, and introduced me to a lifelong hobby. I hope you will recognize yourselves and your important work in this book. The birders I met at the 2015 National Audubon Convention took the time to talk to me on a busy weekend at an important conference that takes place only every other year, and the birders I met at the 2016 World Series of Birding talked to me while physically exhausted, after spending a full twenty-four hours out in the elements birding. The Bedford Audubon Society, Mearns Bird Club, and Rockland Audubon Society allowed me to present my research at their meetings, and I received valuable feedback from them, which I was able to incorporate into the writing of this book. Karen D'Alessandri of Rockland Audubon Society read and provided feedback on one of my chapters in progress. I look forward to sharing this work with birders and to continuing the conversation at chapter meetings and other conferences in the future.

Peter Mickulas of Rutgers University Press believed in this project from the beginning, and his support helped bring the book to fruition. His in-depth understanding of the social and natural worlds was crucial for this project, and he always provided quick and helpful responses to my questions. I'm so pleased the book found a home in the Nature, Society, and Culture series edited by Scott Frickel. Lisa Jean Moore and Leslie Irvine provided especially useful reviews for Rutgers, which I appreciated in addition to their mentorship for my work more generally. Laura Portwood-Stacer of Manuscript Works helped me whittle this down to a manageable size, and I like to think her feedback will permanently affect my writing for the better. Thanks to Daryl Brower of Rutgers University Press and Mary Ribesky of Westchester Publishing Services for their work on the book production, and to Amron Gravett of Wild Clover Book Services for the indexing.

While I wrote most of this book at my desk watching birds feed on the mulberry tree in my backyard, I was fortunate to spend part of the summer of 2017 at Camp Muse in Maine, through the Shin Pond Summer Retreat Program of the Humane Society of the United States. Special thanks go to Bernie Unti of HSUS for organizing the program, and to the Wisemans for use of their property. Funding for this project came from Manhattanville College's summer research stipend and a research sabbatical. Colleagues at Manhattanville College also provided positive, helpful, and most importantly, interdisciplinary feedback at the faculty lecture series and the research blitz.

I had the opportunity to present my analyses to various audiences at different points throughout my data collection and writing, and this work greatly benefited from the feedback I received from other researchers in sociology and in other disciplines. The interdisciplinary group of scholars at the 2016 Fellow Travelers Conference at Wesleyan University, sponsored by the Animals & Society Institute, were among the first to hear about this work. The feedback from scholars in history, literature, education, philosophy, psychology, and more helped me strengthen my arguments as I aimed to reach scholars in multiple disciplines. The Animals & Society section of the American Sociological Association has provided a supportive disciplinary home for human-animal studies, and my work has greatly benefited from section members' comments on my conference presentations on this project.

Parts of chapter 6 originally appeared as "Birding, Citizen Science, and Wildlife Conservation in Sociological Perspective," *Society & Animals* 26 (2018): 130–147. Susan Bullers, David R. Johnson, and the anonymous reviewers for the journal helped me deepen and refine my analyses for the article, and their comments further helped me develop the book chapter beyond the original article's scope. The editors of the journal's special issue on wildlife conservation, Monica Ogra and Julie Urbanik, further supported this project with their expert advice and enthusiasm.

I feel like I've been talking my friends' and colleagues' ears off about birds for the past three-plus years (speaking of ears, did you know birds have ear holes on the side of their head that are covered with special feathers called auriculars that are designed to funnel sound?), and I greatly appreciate their encouragement and friendship. Heather Macpherson Parrott (and her family) were among the first I told about this new project, and our discussions on Teddy Roosevelt's ties to Oyster Bay and visits to the eagle's nest provided much-needed writing breaks to actually go outside and look at birds. Michael

Ramirez and Ross Haenfler helpfully talked through a variety of publishing strategies with me; Jeff Kidder, Nayma Qayum, Ross Collin, and Jessica Greenebaum donated their time and expertise to read chapters in progress; and my writing group of Meghan Freeman, Caralyn Bialo, and Nayma Qayum provided steadfast support throughout this process.

My mother, Ann Cherry, now sends me bird articles alongside vegan articles, and my father, Hank Cherry, and I split a binocular–spotting scope pair, his for scouting fields and mine for watching birds. My mother- and father-in-law, Pam and Dave Saunders, and their interest in birds, really got me thinking about starting this project. Special thanks to Dave for the use of his beautiful bird photographs in this book. My husband, Anthony Saunders, has supported this project in so many ways, and I can only hope that I support his artistic endeavors in kind. From his electroacoustic piece "Grasshopper Sparrow" to his love of Mourning Doves, and with his perfect echoic memory, he'll be a birder yet. I'm so glad we took the Fløibanen funicular to Mount Fløyen in Bergen and stared at that European Robin in the rain.

NOTES

INTRODUCTION

1. Audubon, *Audubon and His Journals*, 470–471.
2. When Charles Darwin published *On the Origin of Species* in 1859, he theorized extinction as part of the process of evolution. Around the same time, naturalists were trying to decipher recently unearthed dinosaur skeletons. However, much of the discussion around evolution and extinction at that time involved natural selection, and not anthropogenic causes of extinction, like overhunting.
3. Audubon, *Audubon and His Journals*, 457.
4. Conniff, *Species Seekers*, 25.
5. Conniff, 146–147.
6. Barrow, *Passion for Birds*, 118.
7. Quoted in Price, *Flight Maps*, 82.
8. Price, 99.
9. Barrow, *Passion for Birds*, 167.
10. Many of these stereotypes are borne of real life. Competitive birding is an element of birding. But as I make clear in this chapter and beyond, I'm studying birders who are not competitive and who are interested in recreation and conservation. For an overview of competitive birders and other types of birders, see McFarlane, "Specialization and Motivations of Birdwatchers."
11. As one example of a big year memoir, see Hayward, *Lost among the Birds*. I describe the practice of traveling to see rare birds later in the book, but for one example, see this story of British birders chartering private flights to the coast of Scotland to see a rare (to them) female Red-winged Blackbird: "Orkney Twitchers Claim European First with Red-Winged Blackbird," BBC News, 2017, https://www.bbc.com/news/uk-scotland-north-east-orkney-shetland-39763010.
12. The World Series of Birding actually has a conservation mission and is sponsored by New Jersey Audubon. For the satirical *Daily Show* clip, see "The World Series of Birding," The Daily Show with Jon Stewart, 2000, http://www.cc.com/video-clips/cxiky4/the-daily-show-with-jon-stewart-the-world-series-of-birding.
13. U.S. Fish & Wildlife Service, "Birding in the United States."
14. As cited in Robinson, "Relative Prevalence of African Americans among Bird Watchers."
15. "Birding in the United States."
16. Scott and Thigpen, "Understanding the Birder as Tourist."
17. I discuss this phenomenon in more detail in several chapters in the book.
18. Lanham, *Home Place*.

19. After the success of the Let's Go Birding Together initiative in 2016, several other Audubon chapters organized their own LGBT walks, such as this one in 2018 in New York City: Dominic Arenas, "'Let's Go Birding Together' Creates a Dedicated Space for LGBTQ Bird Lovers," Audubon, 2018, https://www.audubon.org/news/lets-go-birding-together-creates-dedicated-space-lgbtq-bird-lovers.

20. Human-animal studies scholars like to remind readers that people are mammals by using the terms "human and nonhuman animals." Of course, symbolic boundaries between people and animals have also been used to dehumanize peoples, and thus we must be careful not to replicate those forms of oppression when using those terms. For more information, see Kim, "Moral Extensionism or Racist Exploitation?"

21. Kalof and Fitzgerald, "Reading the Trophy"; Eliason, "Statewide Examination of Hunting and Trophy Nonhuman Animals"; Colomy and Granfield, "Losing Samson"; Hays, "Lie of the Lion."

22. Bayma, "Rational Myth Making and Environment Shaping"; Grazian, *American Zoo*; Jerolmack, *Global Pigeon*.

23. Moore, *Catch and Release*.

24. U.S. Fish and Wildlife Service, "2011 National Survey of Fishing, Hunting, and Wildlife-Associated Recreation."

25. Audubon, *Audubon and His Journals*, 354.

26. Zerubavel, *Hidden in Plain Sight*.

27. Simmel, "Metropolis and Mental Life."

28. Cronon, "Trouble with Wilderness."

29. The birders I interviewed all use the terms "birder" and "birding," rather than "bird watcher" and "bird watching," to describe their identity and their practices. I follow their lead and use their terminology throughout the book, and I explain this in more detail in chapters 1 and 2.

30. Feminist research methods generally describe conducting research not only about but also for the groups being studied. For my work, this meant collaborating with birding groups, donating my time and expertise where needed, such as website design and social media outreach.

31. Fine, "Small Groups and Culture Creation."

32. Starting in July 2015, I conducted thirty in-depth interviews with birders, which I recorded and transcribed. Most of the interviews lasted seventy to ninety minutes, but with a range of thirty minutes to three hours. I ended each interview with an open-ended question so interviewees could address any other aspects of birding that was important to them and that we didn't discuss during the interview.

My interview participants were typically white women, highly educated, and near retirement age. I did not inquire about race, but all but one of my interviewees presented as white. I likewise did not inquire about gender identity, but twenty-two presented as women and eight as men. The group was highly educated, with ten holding master's degrees and five with doctoral degrees. Half of my interviewees held a degree in the natural sciences, and six worked in or retired from a natural science field. Interviewees' ages ranged from eighteen to seventy-nine, with most in their fifties and sixties. They had all been birders for 4–42 years, with an average length of time birding of 23.7 years. I analyzed my transcribed interviews and field notes using the qualitative data analysis software MaxQDA, following a grounded theory approach.

I first used the themes that emerged from my fieldwork to inform certain interview questions. Then, the iterative process of going back and forth between theory and data informed the conception of some of the chapters in this book. What the methodologists Anselm Strauss and Juliet Corbin call "conceptual ordering" helped me derive my interview questions, and what they call "theorizing" led to my discrete chapters, and especially to my development of the concept of the naturalist gaze. My process of constructing explanatory schema was aided by my staying in the field throughout the duration of this entire process, even throughout the writing of this book. See Anselm Strauss and Juliet M. Corbin, *Basics of Qualitative Research: Techniques and Procedures for Developing Grounded Theory* (London: Sage, 1998).

33. Berger, "Why Look at Animals?"

CHAPTER 1 BECOMING A BIRDER

1. Zerubavel, *Hidden in Plain Sight.*
2. Zerubavel's introduction to this book recounts the story of a wilderness coach: "I remember taking one of the wilderness classes I teach out for a walk. We passed a dozen deer, two foxes, one cottontail, six groundhogs, a myriad of birds, insects and other creatures. Nobody noticed even one of them. When I went through the list, the students were angry at themselves: how could they have missed so much?" Brown, "Fill Your Senses," cited in Zerubavel, *Hidden in Plain Sight*, 1.
3. Goodwin, "Professional Vision," cited in Zerubavel, *Hidden in Plain Sight*, 56.
4. Moore, "Birding: Recent Experiences."
5. Zerubavel, *Hidden in Plain Sight*, 11.
6. Zerubavel, 19.
7. Zerubavel, 20.
8. Carson, *Silent Spring.*
9. Zerubavel, *Hidden in Plain Sight*, 23.
10. For a discussion of this dubbing, and for an example of the Bald Eagle's cry, see: Jessica Robinson, "Bald Eagle: A Mighty Symbol, With A Not-So-Mighty Voice," NPR, 2012, https://www.npr.org/templates/story/story.php?storyId=156187375?storyId=156187375.
11. On identifying bird sounds as a way to help solve crimes, see this Guardian story: Stephen Moss, "Can Birdsong Help to Solve Crime?," The Guardian, 2014, https://www.theguardian.com/theguardian/shortcuts/2014/sep/24/can-birdsong-help-to-solve-crime.
12. On blind birders teaching others to bird by ear, see Pat Leonard, "Blind Birders Count By Ear," All About Birds, The Cornell Lab of Ornithology, 2012, https://www.allaboutbirds.org/blind-birders-count-by-ear/. On blind birders forming teams for

birding competitions, see Ralph Blumenthal, "For a Few Birders in Texas, The Victory Is in the Trill," *New York Times*, April 13, 2014.

13. Zerubavel, *Hidden in Plain Sight*, 30.
14. Zerubavel, 32.
15. The sociologist Gary Alan Fine also found that mushroomers identify the mushroom species they find, in part because they will consume them. Fine, *Morel Tales*.
16. I discuss birders' participation in citizen science projects in more detail in chapter 6.
17. Harvey, "Space for Culture and Cognition."
18. A "big year" refers to a year in which a birder attempts to observe the most species possible, and it represents a more competitive aspect to birding than the conservation side that I focus on in this book. Noah Strycker published a book based on his record-breaking big year: *Birding without Borders: An Obsession, a Quest, and the Biggest Year in the World*.
19. On observation in zoos, see Bayma, "Rational Myth Making and Environment Shaping"; Grazian, *American Zoo*.
20. See guideline 6 here: Dave McClain, "Group Rules/Guidelines," Facebook, 2017, https://www.facebook.com/notes/the-facebook-bird-id-group-of-the-world/group-rulesguidelines/1417035835028266.
21. Shepherd, "Classification, Cognition and Context."
22. On deliberative cognition and automatic schemata, see both DiMaggio, "Culture and Cognition," and Cerulo, "Mining the Intersections of Cognitive Sociology and Neuroscience."
23. Shepherd, "Classification, Cognition and Context."
24. Lewis, *Feathery Tribe*.
25. The American Ornithologists' Union described the name change as a desire to "conform to worldwide use" and not to avoid an offensive, race-based name: "Forty-Second Supplement to the American Ornithologists' Union *Check-List of North American Birds*," *The Auk* 117, no. 3 (2000): 847–58, https://doi.org/https://doi.org/10.1642/0004-8038(2000)117[0847:FSSTTA]2.0.CO;2.
26. Examples taken from the table of contents of Peterson, *Peterson Field Guide to Birds of Eastern and Central North America*.
27. On lumping and splitting, see Zerubavel, *Fine Line*, 27–28.
28. Peterson, *Peterson Field Guide to Birds of Eastern and Central North America*, 1.
29. I discuss the importance of appreciating common birds in more detail in chapter 3.
30. Mohr, "Measuring Meaning Structures."
31. Again, I discuss the importance of appreciating common birds in more detail in chapter 3.
32. Zerubavel, *Fine Line*, 35–36.
33. Zerubavel, 115.
34. Csikszentmihalyi, *Flow: The Psychology of Optimal Experience*.
35. Csikszentmihalyi, 71.
36. Csikszentmihalyi, 48.
37. Csikszentmihalyi, 54.

38. On birding and stress reduction, see Cox et al., "Doses of Neighborhood Nature." On how birding with skills reduces stress even more, see Cox and Gaston, "Likeability of Garden Birds."

39. Cox and Gaston, "Likeability of Garden Birds," 53, 62.

40. See Eric Moore, "Lots to See on a Quick Birding Trip to Florida," The Daily Courier, 2017, https://www.dcourier.com/news/2017/feb/23/lots-see-quick-birding-trip-florida/.

41. Csikszentmihalyi, *Flow: The Psychology of Optimal Experience*, 59.

42. On forest therapy, see "Home," Association of Nature & Forest Therapy Guides & Programs, 2019, https://www.natureandforesttherapy.org.

43. "The Practice of Forest Therapy," Association of Nature & Forest Therapy Guides & Programs, 2019, https://www.natureandforesttherapy.org/about/practice.

44. Susanna Curtin also found that people engaging in wildlife tourism were able to achieve a flow state through their concentration on finding wildlife. See Curtin, "Wildlife Tourism."

CHAPTER 2 THE NATURALIST GAZE

1. Beardsworth and Bryman, "Wild Animal in Late Modernity"; Kalof, *Looking at Animals in Human History*; Turner, *Abstract Wild*.

2. Foucault, *Discipline and Punish*; Cherry, "'Pig That Therefore I Am.'".

3. Franklin, *Animals and Modern Culture*.

4. Urry, *Tourist Gaze*; Beardsworth and Bryman, "Wild Animal in Late Modernity."

5. Berger, "Why Look at Animals?," 6.

6. Merriam, *Birds through an Opera-Glass*; Grant, *Our Common Birds and How to Know Them*.

7. Benjamin, *Illuminations: Essays and Reflections*.

8. Benjamin, 223.

9. Turner, *Abstract Wild*, 15.

10. Benjamin, *Illuminations: Essays and Reflections*, 222.

11. Benjamin, 224.

12. Benjamin, 225.

13. Csikszentmihalyi, *Flow: The Psychology of Optimal Experience*.

14. Fine, *Morel Tales*.

15. A "life bird" or "lifer" refers to the first time a birder sees a particular species of bird in their lifetime.

16. On ravers, see Thornton, *Club Cultures*. On drug smugglers, see Adler, *Wheeling and Dealing*.

17. On mushroomers, see Fine, *Morel Tales*. On surfers, see Stranger, *Surfing Life*.

18. Čapek, "Of Time, Space, and Birds."

19. "American Birding Association Code of Birding Ethics," American Birding Association, 2019, http://listing.aba.org/ethics/.

20. For the National Audubon Society's guide to ethical bird photography, see "Audubon's Guide to Ethical Bird Photography," Audubon, 2019, https://www.audubon.org/get-outside/audubons-guide-ethical-bird-photography.
21. Gordon, "Institutional and Impulsive Orientations in Selectively Appropriating Emotions to Self."
22. "World Series of Birding Rules of Competition," New Jersey Audubon, 2017, http://wsb_new.blueskysweet.com/uploads/2017_Exhibit_B-Rules.pdf.
23. Hill, *Ivorybill Hunters*, 39.

CHAPTER 3 COMMON BIRDS AND THE SOCIAL CONSTRUCTION OF NATURE

1. Hayward, *Lost among the Birds*, 163–164.
2. The full checklist can be found here: "Birds of Van Cortlandt Park," Van Cortlandt Park Conservancy, accessed February 20, 2019, http://nycphantom.com/journal/wp-content/uploads/2013/06/VanCortLandtPark-Birds-Checklist.pdf.
3. Cronon, "Trouble with Wilderness," 88.
4. The most popular origin story that walk leaders shared was of the European Starling, brought to the United States by a man who wanted to bring all of the birds mentioned in Shakespeare's works to the United States. I explain this origin story in great detail in chapter 5.
5. Rob Dunn, "The Story of the Most Common Bird in the World," Smithsonian.com, 2012, https://www.smithsonianmag.com/science-nature/the-story-of-the-most-common-bird-in-the-world-113046500/.
6. Of course, the British settlers did not use Native American names for birds, such as the Cherokee word for blackbird, *gv-ne-ga tsi-s-qua*.
7. Mitchell, "Bard's Bird," 177.
8. I discuss birders' evaluation of cowbirds' parasitism in great detail in chapter 5.
9. "Red-Winged Blackbird Identification," All About Birds, The Cornell Lab of Ornithology, accessed February 20, 2019, https://www.allaboutbirds.org/guide/Red-winged_Blackbird/id.
10. Alexandra Topping, "Birdwatchers Flock to Orkney to Catch Glimpse of American Blackbird," The Guardian, 2017, https://www.theguardian.com/environment/2017/may/01/birdwatchers-flock-orkney-catch-glimpse-american-red-winged-blackbird.
11. Tovey, "Theorising Nature and Society in Sociology."
12. On farmed animals, see Buller, "Individuation, the Mass and Farm Animals." On wildlife, see Herda-Rapp and Marotz, "Contested Meanings."
13. For an early example of a sociologist of human-animal studies discussing animal selfhood, see Irvine, *If You Tame Me*.
14. Herda-Rapp and Marotz, "Contested Meanings."
15. Gary Alan Fine writes, "Just as birders speak of LBJs ('little brown jobbies'), rockhounds speak of 'rock rubbish,' and fishers speak of 'trash fish,' mushroomers have their

own terms for mushrooms that are uninteresting." Fine, *Morel Tales*, 70. And in the introduction to their edited volume *Trash Birds*, Kelsi Nagy and Phillip David Johnson II write, "'Environmentally minded' citizens, such as birders, often refer to common birds, such as pigeons and Canadian geese, as 'garbage' or 'trash' birds, exposing a deeper prejudice invoked by some ubiquitous birds even in so-called nature lovers." Nagy and Johnson, "Introduction," 9. In contrast to these anecdotal examples, I contend that conservation-minded birders demonstrate a vast appreciation for common birds, and that any denigration that casual observers might notice comes in the form of joking.

16. Kalof, Zammit-Lucia, and Kelly, "Meaning of Animal Portraiture in a Museum Setting."

17. Fine, *Morel Tales*; Nagy and Johnson, "Introduction."

18. Čapek, "Surface Tension."

19. On moral shocks, see Jasper and Poulsen, "Recruiting Strangers and Friends."

20. Birders at an observatory in Quebec recently spotted 700,000 warblers in one day, in a feat that even the *New York Times* reported. Gorman, "A River of Warblers."

21. Fine, *Morel Tales*, 71.

22. Greider and Garkovich, "Landscapes: The Social Construction of Nature and the Environment," 1.

23. On the eco-self, see Čapek, "Surface Tension."

24. Cronon, "Trouble with Wilderness," 80.

25. Cronon, 86.

26. Cronon, 87.

27. "Urban and Rural Population for the U.S. and All States: 1900–2000," Iowa State Data Center, accessed February 12, 2019, https://www.iowadatacenter.org/datatables/UnitedStates/urusstpop19002000.pdf.

28. "Rural America at a Glance: 2016 Edition," United States Department of Agriculture, 2016, https://www.ers.usda.gov/webdocs/publications/80894/eib-162.pdf?v=42684.

29. Berger, "Why Look at Animals?," 26.

30. Anthony Leiserowitz et al., "Global Warming's Six Americas," Yale Program on Climate Change Communication, 2016, http://climatecommunication.yale.edu/about/projects/global-warmings-six-americas/.

CHAPTER 4 WILDERNESS, WILDNESS, AND MOBILITY

1. We can see this in the difference between wild birds and domesticated flightless birds that are raised for farming. Domesticated flightless birds such as chickens, turkeys, emus, and ostriches do not exemplify wilderness and wildness in the same ways, because they do not freely fly from place to place. They are completely controlled by humans.

2. Csikszentmihalyi, *Flow: The Psychology of Optimal Experience*.

3. Csikszentmihalyi, 71.

4. On whale-watching tours as guided narration, see Grahame, "Looking at Whales."

5. Deeply interested in words and their meanings, the poet and environmental literature professor Gary Snyder explores the etymology and historical definitions of the terms.

On wilderness, Snyder finds several meanings, including "a large area of wild land," "a wasteland," "a space of sea or air," "a place of danger or difficulty," "the world as contrasted with heaven," or "a place of abundance." Snyder, *Practice of the Wild*, 11. On wildness, Snyder notes that dictionary definitions tend to define "wild" by what, from a human perspective, something is not: undomesticated, unruly, uninhabited, uncultivated (9). Turning this perspective around, Snyder creates new definitions of what wild is: "Of animals—free agents, each with its own endowments, living within natural systems. Of plants—self-propagating, self-maintaining, flourishing in accord with innate qualities. Of land—a place where the original and potential flora and fauna are intact and in full interaction and the landforms are entirely the results of nonhuman forces" (9–10). Given these understandings, Snyder then notes how wilderness and wildness cohabitate: "Wilderness is a place where the wild potential is fully expressed, a diversity of living and nonliving beings flourishing according to their own sorts of order. In ecology we speak of 'wild systems.' When an ecosystem is fully functioning, all the members are present at the assembly. To speak of wilderness is to speak of wholeness" (12).

6. Alex Furuya, "Are Birds Nesting Earlier So Their Chicks Don't Overheat?," Audubon, 2017, https://www.audubon.org/news/are-birds-nesting-earlier-so-their-chicks-dont-overheat.

7. The Wilderness Act of 1964 defines wilderness as "an area where the earth and its community of life are untrammeled by man, where man himself is a visitor who does not remain" ("The Wilderness Act of 1964," United States Department of Justice, 2015, https://www.justice.gov/enrd/wilderness-act-1964.). Many scholars of wilderness, especially J. Baird Callicott and William Cronon, critique this way of conceiving of wilderness, known as the "received wilderness idea." I agree with these authors that we can critique this idea of wilderness while working to protect wilderness areas. For more information, see Callicott, "Critique of and Alternative to the Wilderness Idea"; Cronon, "Trouble with Wilderness."

8. Enright, "Why the Rhinoceros Doesn't Talk," 110.

9. Less than 5 percent of the land in the United States is protected wilderness areas. I discuss this phenomenon in more detail in chapter 7.

10. Enright, "Why the Rhinoceros Doesn't Talk," 122.

11. Snyder, *Practice of the Wild*, 11.

12. Cronon, "Trouble with Wilderness," 89.

13. Cronon, 87.

14. Cronon, 89.

15. The term "speciesism" generally describes the victimization of animals as an oppression akin to other socially constructed hierarchies and oppressions such as racism and sexism. For more information, see Shapiro, "Animal Rights versus Humanism."

16. Snyder, *Practice of the Wild*, 6.

17. Enright, "Why the Rhinoceros Doesn't Talk."

18. "When it comes to wild others, like Genet and wild land, which must continue to exist or be posited as existing in contrast and opposition to imperial power if power is to save its reality principle, the risk of the real is somewhat amplified. In order to do the job

of preserving its reality principle, and in spite of its need to simulate or define the other according to its own models, the imperium must leave at least enough otherness intact to *maintain the glance of the other*. The other must be able to cast its glance at the imperial enterprise to preserve the meaning of that enterprise, to legitimate its purpose of bringing law and order to wild chaos, and to threaten those who might question its good intentions and overall beneficence. There must remain at least some vestiges of wildness to be kept at bay." Birch, "Incarceration of Wilderness," 19.

19. Armbruster, "Into the Wild," 766.

20. I discuss birds and science education in more detail in chapter 7.

21. Cronon, "Trouble with Wilderness," 85.

22. Snyder, *Practice of the Wild*, 14.

CHAPTER 5 GOOD BIRDS, BAD BIRDS, AND ANIMAL AGENCY

1. Arluke and Sanders, *Regarding Animals*, 169.

2. Arluke and Sanders, 169.

3. Arluke and Sanders, 170.

4. Martin, "When Sharks (Don't) Attack."

5. Irvine, *If You Tame Me*, 128.

6. Moore and Kosut, *Buzz: Urban Beekeeping*, 106.

7. The sociologist Richie Nimmo contends that "agency" is inherently anthropocentric and is a product of the human-animal divide: "To insist upon the agential status of nonhuman animals is effectively to anthropomorphize them, and is therefore no solution." Nimmo, *Milk, Modernity and the Making of the Human, Purifying the Social*, 41. Instead, Nimmo suggests that scholars conceive of animal agency in terms of movement, or flows, as this "foregrounds a notion of animal agency as something more fluid, which permeates the ensemble of social and material movements of which the animals are a part." Nimmo, "Bovine Mobilities and Vital Movements," 59. To Nimmo, the flows of milk represent a vital element of cows' agency. In this chapter, I argue that birds' flight, their mobility, represents a key aspect of birds' agency. But as wild animals, birds exemplify more of Irvine's and Moore's definitions of agency, as they maintain some of their free will, including their ability to persist, resist, and make choices on how to act and interact within the structures of human-dominated landscapes.

8. On companion animals, see Irvine, *If You Tame Me*. On lab animals, see Herzog, "Moral Status of Mice." On farmed animals, see Nimmo, *Milk, Modernity and the Making of the Human*; Moore and Kosut, *Buzz: Urban Beekeeping*. On animals kept in captivity for entertainment in zoos, see Grazian, *American Zoo*.

9. Martin, "When Sharks (Don't) Attack."

10. The Christmas Bird Count, discussed in detail in chapter 6, was itself explicitly patterned after Christmas Day hunts, where the goal was to kill the largest number of species of birds. The Christmas Bird Count, in contrast, sought to observe the largest number of species of birds, without killing them.

11. Irvine, *If You Tame Me*, 81.
12. Jerolmack, *Global Pigeon*, 107.
13. Conniff, *Species Seekers*.
14. Arluke and Sanders, *Regarding Animals*, 171–173.
15. Van Dooren, *Flight Ways*.
16. Van Dooren, 51.
17. Van Dooren, 53. See also Elliot Hannon, "Vanishing Vultures: A Grave Matter for India's Parsis," NPR, 2012, https://www.npr.org/2012/09/05/160401322/vanishing-vultures-a-grave-matter-for-indias-parsis.
18. Van Dooren, *Flight Ways*, 54.
19. Jerolmack, *Global Pigeon*.
20. Peterson, *Peterson Field Guide to Birds of Eastern and Central North America*, 228.
21. Ritvo, "Going Forth and Multiplying," 2n2.
22. "American Acclimatization Society," *New York Times*, November 15, 1877.
23. "American Acclimatization Society."
24. Fine and Christoforides, "Dirty Birds, Filthy Immigrants, and the English Sparrow War," 378.
25. Gary Alan Fine and Lazaros Christoforides studied people's reactions to the English Sparrow (now House Sparrow) as symbolizing people's reactions to immigrants during the "Great English Sparrow War" of 1850–1890. Once the English Sparrows began to thrive in the United States, they came to be hated by birders and naturalists. Fine and Christoforides argue that the hatred came from viewing the birds as moral symbols: "(1) they were foreign 'immigrants,' (2) they attacked native ('American') birds, (3) their character (cleanliness, noise, sexual habits) was seen as disreputable, and (4) they needed to be controlled as a foreign enemy." Fine and Christoforides, "Dirty Birds, Filthy Immigrants, and the English Sparrow War," 381. Fine and Christoforides also describe how nationalism could be seen in the birding checklists from the American Ornithologists' Union, which debated whether to include the "foreign" English Sparrows on its checklists. Given the nationalistic motivation for the union's and birders' debates on English Sparrows, I would not consider their evaluations of the sparrows to come from the naturalist gaze.
26. Allison Wells, "Impact of House Sparrow and Other Invasive Bird Species Being Monitored by Volunteers in Cornell Lab of Ornithology's Birdhouse Network," Cornell Chronicle, 2004, http://news.cornell.edu/stories/2004/04/citizens-asked-monitor-impact-invasive-bird-species.
27. See "Farmed Animal Statistics: Millions of Wildlife Killed Annually for U.S. Farmers," Farmed Animal Watch, 2005, http://www.farmedanimal.org/faw/faw5-35.htm, for a full list of animals killed by USDA Wildlife Services Division in 2004. Subsequent reports are unavailable. Note that Pigeons, Doves, Crows, Ravens, Manikins, and the generic "Blackbird" all account for more animals killed than Sparrows.
28. European Starlings and House Sparrows exemplify liminal animals that survive on anthropogenic food sources. In their research on the depiction of liminal animals in *National Geographic*, the sociologists Linda Kalof, Cameron Whitley, Steven Vrla, and

Jessica Bell Rizzolo found differences between birds that were invited to feed at feeders or live in bird houses and uninvited birds that "took up residence" in spaces like barns. Liminal animals, which were not fully wild and not fully domesticated, searched for anthropogenic food sources. Kalof et al., "Anthropogenic Food Sources in the Co-existence of Humans with Liminal Animals in Northern Environments."

29. These evaluations echo the findings of the sociologist Linda Kalof and the education scholar Ramona Fruja Amthor, who studied the cultural representations of "problem animals" as they were depicted in *National Geographic*. Two of the three main themes they found in these cultural representations blamed the animals—problem animals were "dangerous and disruptive to humans and their property," and they were "dangerous and disruptive to the natural world." However, their third theme blamed humans, who were also depicted as "dangerous and disruptive to the natural world." Kalof and Amthor, "Cultural Representations of Problem Animals in National Geographic."

30. In some ways, this judgment resembles people's judgments against Pit Bull dogs, especially when Pit Bull advocates remind others that Pit Bulls are loyal and harmless, and dangerous only if trained to be aggressive. But in other ways, this judgment differs, since it's based on birders knowing birds' instinctual behaviors because of scientific research, whereas the judgments of anti–Pit Bull people are based on anecdotal evidence and emotions.

31. Arluke and Sanders, *Regarding Animals*, 180–186.

32. Pit Bull terriers historically were known as "nanny dogs" because of their gentle and protective nature. Now, people stigmatize them as dangerous when they pass temperament tests at higher rates than breeds typically viewed as less "dangerous," such as Cocker Spaniels, Dachshunds, and Golden Retrievers, according to the American Temperament Test Society: "ATTS Breed Statistics," American Temperance Test Society, 2017, https://atts.org/breed-statistics/statistics-page8/. For more information on Pit Bulls and breed stigma, see Twining, Patronek, and Arluke, "Managing the Stigma of Outlaw Breeds."

33. "Brown-headed Cowbirds," NestWatch, Cornell Lab of Ornithology, accessed February 12, 2019, https://nestwatch.org/learn/general-bird-nest-info/brown-headed-cowbirds/.

34. "Brown-headed Cowbirds," NestWatch.

35. "Brown-headed Cowbirds," NestWatch.

36. Howell, Dijak, and Thompson, "Landscape Context and Selection for Forest Edge by Breeding Brown-Headed Cowbirds."

37. Amy Lewis, "Is It Okay to Remove Cowbird Eggs From Host Nests?," Audubon, 2018, https://www.audubon.org/news/is-it-okay-remove-cowbird-eggs-host-nests?

38. Irvine, *If You Tame Me*, 128.

39. Moore and Kosut, *Buzz: Urban Beekeeping*, 206.

40. Louder et al., "Generalist Brood Parasite Modifies Use of a Host in Response to Reproductive Success."

41. Martin, "When Sharks (Don't) Attack."

CHAPTER 6 BIRDING AND CITIZEN SCIENCE

1. Barrow, *Passion for Birds.*
2. Quoted in Dunlap, *In the Field,* 83.
3. Barrow, *Passion for Birds,* 168.
4. Schaffner, "Environmental Sporting."
5. Schaffner, *Binocular Vision.*
6. I use the term "citizen science" to describe science projects conducted by amateurs, although the term was coined in the 1990s. The practice of citizen science has a much longer history, and the fields of astronomy and ornithology produced the first citizen science projects. In 1749, amateur ornithologists began collecting data on bird migration in Finland, and in 1874 the British Transit of Venus project employed amateur astronomers around the globe to measure the earth's distance to the sun. On Finnish amateur ornithologists, see Greenwood, "Citizens, Science and Bird Conservation." On the Transit of Venus project, see Ratcliff, *Transit of Venus Enterprise in Victorian Britain.*
7. Barrow, *Passion for Birds.*
8. Lewis, *Feathery Tribe.*
9. Barrow, *Passion for Birds.*
10. Dunlap, *In the Field,* 198.
11. "House Finch Disease Survey."
12. Dickinson, Zuckerberg, and Bonter, "Citizen Science as an Ecological Research Tool."
13. Weidensaul, *Of a Feather,* 4.
14. Moore and Kosut, *Buzz: Urban Beekeeping,* 57–58.
15. Scroggins, "Ignoring Ignorance," 207.
16. Epstein, "Construction of Lay Expertise."
17. Wohlsen, *Biopunk: Solving Biotech's Biggest Problems.*
18. Mills, *White Collar.*
19. Gerth and Mills, *Character and Social Structure.*
20. Kidder, "'It's the Job That I Love.'"
21. Sullivan et al., "EBird: A Citizen-Based Bird Observation Network," 2285.
22. Freidson, *Profession of Medicine.*
23. Wynne, "Misunderstood Misunderstanding."
24. Epstein, "Construction of Lay Expertise," 411.
25. On conservation and investigation goals, see Wiggins and Crowston, "From Conservation to Crowdsourcing."
26. On crowdsourcing data, see Law et al., "Crowdsourcing as a Tool for Research."
27. Riesch and Potter, "Citizen Science as Seen by Scientists."
28. Law et al., "Crowdsourcing as a Tool for Research."
29. Shirk, "Push the Edge of Science Forward."
30. Aronson, "Science as a Claims-Making Activity."
31. Burgess et al., "Science of Citizen Science."
32. Burgess et al., 6.

33. Hilton-Taylor, "Evolving IUCN Red List Needs to Be Both Innovative and Rigorous."
34. Kennett, "Participatory Monitoring and Management, and Citizen Science."
35. Wiggins et al., "Mechanisms for Data Quality and Validation in Citizen Science."
36. Mulder et al., "Citizen Science.".
37. Abbott, *System of Professions*; Freidson, *Profession of Medicine*.
38. For a list of publications using eBird data, see: "Publications," eBird, 2019, https://ebird.org/science/publications. For a list of publications using CBC data, see: "Christmas Bird Count Bibliography," Audubon, 2019, https://www.audubon.org/christmas-bird-count-bibliography.
39. On scientific credibility, see Epstein, "Construction of Lay Expertise."
40. Ottinger, "Epistemic Fencelines."
41. Freidson, *Profession of Medicine*; Abbott, *System of Professions*.
42. Wiggins et al., "Mechanisms for Data Quality and Validation in Citizen Science."
43. Collins, *Are We All Scientific Experts Now?*; Wynne, "Misunderstood Misunderstanding."
44. Dunlap, *In the Field*.
45. Boxall and McFarlane, "Human Dimensions of Christmas Bird Counts."

CHAPTER 7 BIRDING AS A CONSERVATION MOVEMENT

1. On the use of bees in the military, see Moore and Kosut, *Buzz: Urban Beekeeping*, 203.
2. Gregory and van Strien, "Wild Bird Indicators."
3. Kat Eschner, "The Story of the Real Canary in the Coal Mine," Smithsonian.com, 2016, https://www.smithsonianmag.com/smart-news/story-real-canary-coal-mine-180961570/.
4. Some miners carried "humane cages" that allowed them to resuscitate their canaries: "Rebecca's Object of the Week: Haldane Canary Cage," National Mining Museum, accessed February 14, 2019, https://nationalminingmuseum.com/rebeccas-object-of-the-week/.
5. Haenfler, Johnson, and Jones, "Lifestyle Movements"; Cherry, "I Was a Teenage Vegan."
6. On supportive social networks and veganism, see Cherry, "Veganism as a Cultural Movement." On "virtuous circles" and green living, see Kennedy, "Rethinking Ecological Citizenship."
7. For more on distancing and the commodity chain, see Carolan, "Unmasking the Commodity Chain."
8. Tidwell, "Low-Carbon Diet."
9. Wynes and Nicholas, "Climate Mitigation Gap."
10. Norgaard, *Living in Denial*.
11. Virginia Morrell, "Meat-Eaters May Speed Worldwide Species Extinction, Study Warns," Science, 2015, http://www.sciencemag.org/news/2015/08/meat-eaters-may-speed-worldwide-species-extinction-study-warns.

12. On the relationship between invasive species and economic development, see Besek and McGee, "Introducing the Ecological Explosion." On the cost of invasive species, see Pimentel, Zuniga, and Morrison, "Update on the Environmental and Economic Costs Associated with Alien-Invasive Species in the United States."

13. As discussed in chapter 2, glass strikes cause the most bird deaths. (See Sibley Guides for more information: "Causes of Bird Mortality," Sibley Guides, 2003, https://www.sibleyguides.com/conservation/causes-of-bird-mortality/.) But mortality rates do not represent future threats, as climate change and habitat loss do. Habitat loss represents the largest threat to birds, for their future viability, as explained by Audubon: Emma Bryce, "Global Study Reveals the Extent of Habitat Fragmentation," Audubon, 2015, https://www.audubon.org/news/global-study-reveals-extent-habitat-fragmentation.

14. Vincent, Hanson, and Bjelopera, "Federal Land Ownership: Overview and Data."

15. Development is the smallest use of privately owned land. Most privately owned land is used for crops, grazing, pastures, or forestry: Cynthia Nickerson and Allison Borchers, "How Is Land in the United States Used? A Focus on Agricultural Land," United States Department of Agriculture, 2012, https://www.ers.usda.gov/amber-waves/2012/march/data-feature-how-is-land-used.

16. Runge et al., "Protected Areas and Global Conservation of Migratory Birds."

17. Runge et al.

18. La Sorte et al., "Global Change and the Distributional Dynamics of Migratory Bird Populations Wintering in Central America."

19. La Sorte et al.

20. Iwamura et al., "Migratory Connectivity Magnifies the Consequences of Habitat Loss from Sea-Level Rise for Shorebird Populations."

21. Iwamura et al.

22. Larson, "New Estimates of Value of Land of the United States."

23. See U.S. Census data on urban and rural populations: "2010 Census Urban and Rural Classification and Urban Area Criteria," United States Census Bureau, accessed February 14, 2019, https://www.census.gov/geo/reference/ua/urban-rural-2010.html.

24. Birders outside of the Audubon Society also recognize the importance of instilling a love of birds, birding, and nature in children. The celebrated birder Victor Emanuel describes his life of birding and his experience founding birding camps for children in Emanuel and Walsh, *One More Warbler*.

25. On zoos and their impression management, see Grazian, *American Zoo*.

26. Russell, "'Everything Has to Die One Day.'"

27. In a separate paper, I argue that birders avoid labeling themselves as "environmentalists," even if their practices and beliefs indicated a much more radical belief system than their self-proclaimed identities. In the interest of space and of maintaining a focus on birders and birding, that argument does not appear in this book.

28. Hatvany, "Imagining Duckland."

29. While Ringer encouraged participants to do "absolutely anything" to spread the word, "absolutely anything" excluded several tactics, such as protesting, direct action, tabling, leafleting, or any kind of theatrical action to bring attention to the needs of wild birds. He didn't say "Dress up like a Kirtland's Warbler and protest outside of the USDA to bring attention to how deforestation for agricultural lands is providing more edge habitat for Brown-headed Cowbirds, whose brood parasitism is endangering Kirtland's Warblers."

30. On "tastes in tactics," see Jasper, *Art of Moral Protest.*

CONCLUSION

1. Audubon, *Audubon and His Journals*, 350.
2. Audubon, 350.
3. Zerubavel, *Hidden in Plain Sight.*
4. Zerubavel, *Fine Line.*
5. "The Migratory Bird Treaty Act makes it illegal to take, possess, import, export, transport, sell, purchase, barter, or offer for sale, purchase, or barter, any migratory bird, or the parts, nests, or eggs of such a bird except under the terms of a valid Federal permit." "Migratory Bird Treaty Act," U.S. Fish & Wildlife Service, 2018, https://www.fws.gov/birds/policies-and-regulations/laws-legislations/migratory-bird-treaty-act.php.
6. Darryl Fears and Dino Grandoni, "The Trump Administration Has Officially Clipped the Wings of the Migratory Bird Treaty Act," The Washington Post, 2018, https://www.washingtonpost.com/news/energy-environment/wp/2018/04/13/the-trump-administration-officially-clipped-the-wings-of-the-migratory-bird-treaty-act/?utm_term=.972d7494f66e.
7. Michael Greshko et al., "A Running List of How President Trump Is Changing Environmental Policy," National Geographic, 2019, https://news.nationalgeographic.com/2017/03/how-trump-is-changing-science-environment/.
8. Wu et al., "Projected Avifaunal Responses to Climate Change across the U.S. National Park System."
9. Wynes and Nicholas, "Climate Mitigation Gap."
10. For the *Audubon* article, see Tidwell, "Low-Carbon Diet." For the Food and Agriculture Organization report, see Steinfeld et al., "Livestock's Long Shadow."
11. Catton and Dunlap, "Environmental Sociology."

BIBLIOGRAPHY

Abbott, Andrew. *The System of Professions: An Essay on the Division of Expert Labor*. Chicago: University of Chicago Press, 1988.

Adler, Patricia A. *Wheeling and Dealing: An Ethnography of an Upper-Level Drug Dealing and Smuggling Community*. New York: Columbia University Press, 1993.

"American Acclimatization Society." *New York Times*, November 15, 1877.

Arluke, Arnold, and Clinton R. Sanders. *Regarding Animals*. Philadelphia: Temple University Press, 1996.

Armbruster, Karla. "Into the Wild: Response, Respect, and the Human Control of Canine Sexuality and Reproduction." *JAC* 30, nos. 3–4 (2010): 755–783.

Aronson, Naomi. "Science as a Claims-Making Activity: Implications for Social Problems Research." In *Studies in the Sociology of Social Problems*, edited by Joseph W. Schneider and John I. Kitsuse, 1–30. Ablex Publishing Corporation, 1984.

Audubon, Maria R. *Audubon and His Journals*. New York: Charles Scribner's Sons, 1897.

Barrow, Mark V. *A Passion for Birds: American Ornithology after Audubon*. Princeton, NJ: Princeton University Press, 1998.

Bayma, Todd. "Rational Myth Making and Environment Shaping: The Transformation of the Zoo." *Sociological Quarterly* 53, no. 1 (2012): 116–141.

Beardsworth, A., and A. Bryman. "The Wild Animal in Late Modernity: The Case of the Disneyization of Zoos." *Tourist Studies* 1, no. 1 (2001): 83–104. https://doi.org/10.1177/146879760100100105.

Benjamin, Walter. *Illuminations: Essays and Reflections*. New York: Harcourt, Brace & World, 1968.

Berger, John. "Why Look at Animals?" In *About Looking*, edited by John Berger, 1–28. New York: Pantheon, 1980.

Besek, Jordan Fox, and Julius Alexander McGee. "Introducing the Ecological Explosion." *International Journal of Sociology* 44, no. 1 (2014): 75–93. https://doi.org/10.2753/IJS0020-7659440105.

Birch, Thomas H. "The Incarceration of Wilderness: Wildness Areas as Prisons." *Environmental Ethics* 12, no. 1 (1990): 3–26.

Boxall, Peter C., and Bonita L. McFarlane. "Human Dimensions of Christmas Bird Counts: Implications for Nonconsumptive Wildlife Recreation Programs." *Wildlife Society Bulletin* 21 (1993): 390–396.

Brown, Tom. "Fill Your Senses, Light Up Your Life." *Reader's Digest*, August 1984.

Buller, Henry. "Individuation, the Mass and Farm Animals." *Theory, Culture & Society* 30, nos. 7/8 (2013): 155–176.

Burgess, H. K., L. B. DeBey, H. E. Froehlich, N. Schmidt, E. J. Theobald, A. K. Ettinger, J. HilleRisLambers, J. Tewksbury, and J. K. Parrish. "The Science of Citizen Science:

Exploring Barriers to Use as a Primary Research Tool." *Biological Conservation* 208 (2017): 113–20. https://doi.org/10.1016/j.biocon.2016.05.014.

Callicott, J. Baird. "A Critique of and Alternative to the Wilderness Idea." *Wild Earth* 4, no. 4 (1994): 54–59.

Čapek, Stella M. "Of Time, Space, and Birds: Cattle Egrets and the Place of the Wild." In *Mad about Wildlife: Looking at Social Conflict over Wildlife*, edited by Ann Herda-Rapp and Theresa Goedeke, 195–222. Leiden, UK: Brill, 2005.

———. "Surface Tension: Boundary Negotiations around Self, Society, and Nature in a Community Debate over Wildlife." *Symbolic Interaction* 29, no. 2 (2006): 157–181.

Carolan, Michael S. "Unmasking the Commodity Chain." *Peace Review* 16, no. 2 (2004): 193–198. https://doi.org/10.1080/1040265042000237734.

Carson, Rachel. *Silent Spring*. New York: Houghton Mifflin, 1962.

Catton, William R, and Riley E Dunlap. "Environmental Sociology: A New Paradigm." *The American Sociologist* 13, no. 1 (1978): 41–49.

Cerulo, Karen A. "Mining the Intersections of Cognitive Sociology and Neuroscience." *Poetics* 38, no. 2 (2010): 115–132. https://doi.org/10.1016/j.poetic.2009.11.005.

Cherry, Elizabeth. "I Was a Teenage Vegan: Motivation and Maintenance of Lifestyle Movements." *Sociological Inquiry* 85, no. 1 (2015): 55–74.

———. "The Pig That Therefore I Am." *Humanity & Society* 40, no. 1 (2016): 64–85. https://doi.org/10.1177/0160597615586620.

———. "Veganism as a Cultural Movement: A Relational Approach." *Social Movement Studies* 5, no. 2 (2006): 155–170. https://doi.org/10.1080/14742830600807543.

Collins, Harry. *Are We All Scientific Experts Now?* Cambridge: Polity, 2014.

Colomy, Paul, and Robert Granfield. "Losing Samson: Nature, Crime, and Boundaries." *Sociological Quarterly* 51, no. 3 (2010): 355–383.

Conniff, Richard. *The Species Seekers: Heroes, Fools, and the Mad Pursuit of Life on Earth*. New York: W.W. Norton, 2011.

Cox, Daniel T. C., and Kevin J. Gaston. "Likeability of Garden Birds: Importance of Species Knowledge and Richness in Connecting People to Nature." *PLoS ONE* 10, no. 11 (2015): e0141505. https://doi.org/https://doi.org/10.1371/journal.pone.0141505.

Cox, Daniel T. C., Danielle F. Shanahan, Hannah L. Hudson, Kate E. Plummer, Gavin M. Siriwarenda, Richard A. Fuller, Karen Anderson, Steven Hancock, and Kevin J. Gaston. "Doses of Neighborhood Nature: The Benefits for Mental Health of Living with Nature." *BioScience* 67, no. 2 (2017): 147–155.

Cronon, William. "The Trouble with Wilderness; or, Getting Back to the Wrong Nature." In *Uncommon Ground: Rethinking the Human Place in Nature*, edited by William Cronon, 69–90. New York: W.W. Norton, 1996.

Csikszentmihalyi, Mihaly. *Flow: The Psychology of Optimal Experience*. New York: Harper Collins, 1990.

Curtin, Susanna. "Wildlife Tourism: The Intangible, Psychological Benefits of Human-Wildlife Encounters." *Current Issues in Tourism* 12, nos. 5–6 (2009): 451–474.

Dickinson, Janis L., Benjamin Zuckerberg, and David N. Bonter. "Citizen Science as an Ecological Research Tool: Challenges and Benefits." *Annual Review of Ecology, Evolution, and Systematics* 41 (2010): 149–172.

DiMaggio, Paul. "Culture and Cognition." *Annual Review of Sociology* 23 (1997): 263–287.

Dunlap, Thomas R. *In the Field, among the Feathered: A History of Birders and Their Guides*. New York: Oxford University Press, 2011.

Eliason, Stephen L. "A Statewide Examination of Hunting and Trophy Nonhuman Animals: Perspectives of Montana Hunters." *Society and Animals* 16 (2008): 256–278. https://doi.org/10.1163/156853008X323402.

Emanuel, Victor, and S. Kirk Walsh. *One More Warbler: A Life with Birds*. Austin: University of Texas Press, 2017.

Enright, Kelly. "Why the Rhinoceros Doesn't Talk: The Cultural Life of a Wild Animal in America." In *Beastly Natures: Animals, Humans, and the Study of History*, edited by Dorothee Brantz, 108–126. Charlottesville: University of Virginia Press, 2010.

Epstein, Steven. "The Construction of Lay Expertise: AIDS Activism and the Forging of Credibility in the Reform of Clinical Trials." *Source: Science, Technology, & Human Values Medical Practices and Science and Technology Studies* 20, no. 4 (1995): 408–437. http://www.jstor.org/stable/689868.

Fine, Gary Alan. *Morel Tales: The Culture of Mushrooming*. Cambridge, MA: Harvard University Press, 1998.

———. "Small Groups and Culture Creation: The Idioculture of Little League Baseball Teams." *American Sociological Review* 44, no. 5 (1979): 733–745.

Fine, Gary Alan, and Lazaros Christoforides. "Dirty Birds, Filthy Immigrants, and the English Sparrow War: Metaphorical Linkage in Constructing Social Problems." *Symbolic Interaction* 14, no. 4 (1991): 375–393. https://doi.org/10.1525/si.1991.14.4.375.

Foucault, Michel. *Discipline and Punish*. Harmondsworth, UK: Penguin, 1995. https://doi.org/10.1073/pnas.0703993104.

Franklin, Adrian. *Animals and Modern Culture: A Sociology of Human-Animal Relations in Modernity*. London: Sage, 1999.

Freidson, Eliot. *Profession of Medicine: A Study of the Sociology of Applied Knowledge*. Chicago: University of Chicago Press, 1970.

Gerth, Hans, and C. Wright Mills. *Character and Social Structure: The Psychology of Social Institutions*. New York: Harcourt, Brace & World, 1953.

Goodwin, Charles. "Professional Vision." *American Anthropologist* 96 (1994): 606–633.

Gordon, Steven L. "Institutional and Impulsive Orientations in Selectively Appropriating Emotions to Self." In *The Sociology of Emotions: Original Essays and Research Papers*, edited by D. D. Franks and E. D. McCarthy, 115–135. Greenwich, CT: JAI Press, 1989.

Gorman, James. "A River of Warblers: 'The Greatest Birding Day of My Life,'" *New York Times*, May 31, 2018, https://www.nytimes.com/2018/05/31/science/warblers-canada-migration.html.

Grahame, Peter R. "Looking at Whales: Narration and the Organization of Visual Experience." *Journal of Contemporary Ethnography* 47, no. 6 (2018): 782–806. https://doi.org/10.1177/0891241618768071.

Grant, John Beveridge. *Our Common Birds and How to Know Them*. New York: C. Scribner's Sons, 1891.

Grazian, David. *American Zoo: A Sociological Safari*. Princeton, NJ: Princeton University Press, 2015.

Greenwood, Jeremy. "Citizens, Science and Bird Conservation." *Journal of Ornithology* 148, no. Supplement 1 (2007): S77–124.

Gregory, Richard D., and Arco van Strien. "Wild Bird Indicators: Using Composite Population Trends of Birds as Measures of Environmental Health." *Ornithological Science* 9, no. 1 (2010): 3–22. https://doi.org/10.2326/osj.9.3.

Greider, Thomas, and Lorraine Garkovich. "Landscapes: The Social Construction of Nature and the Environment." *Rural Sociology* 59, no. 1 (1994): 1–24.

Haenfler, Ross, Brett Johnson, and Ellis Jones. "Lifestyle Movements: Exploring the Intersection of Lifestyle and Social Movements." *Social Movement Studies* 11, no. 1 (2012): 1–20.

Harvey, Daina Cheyenne. "The Space for Culture and Cognition." *Poetics* 38, no. 2 (2010): 184–203. https://doi.org/10.1016/j.poetic.2009.11.009.

Hatvany, Matthew G. "Imagining Duckland: Postnationalism, Waterfowl Migration, and Ecological Commons." *The Canadian Geographer* 61, no. 2 (2017): 224–39. https://doi.org/10.1111/cag.12352.

Hays, Cassie M. "The Lie of the Lion: Racialization of Nature in the Safari Souvenir." *Environmental Sociology* 1, no. 1 (2015): 4–17. https://doi.org/10.1080/23251042.2014.971479.

Hayward, Neil. *Lost among the Birds: Accidentally Finding Myself in One Very Big Year*. New York: Bloomsbury USA, 2016.

Herda-Rapp, Ann, and Karen G. Marotz. "Contested Meanings: The Social Construction of the Mourning Dove in Wisconsin." In *Mad about Wildlife: Looking at Social Conflict over Wildlife*, edited by Ann Herda-Rapp and Theresa L. Goedeke, 73–98. Leiden, The Netherlands: Brill, 2005.

Herzog, Hal. "The Moral Status of Mice." *American Psychologist* 43 (1988): 473–474.

Hill, Geoffrey E. *Ivorybill Hunters: The Search for Proof in a Flooded Wilderness*. New York: Oxford University Press, 2007.

Hilton-Taylor, Craig. "An Evolving IUCN Red List Needs to Be Both Innovative and Rigorous." Mongabay. Accessed August 24, 2017. https://news.mongabay.com/2017/05/an-evolving-iucn-red-list-needs-to-be-both-innovative-and-rigorous/.

"House Finch Disease Survey." Cornell Lab of Ornithology. Accessed January 9, 2017. http://www.birds.cornell.edu/citscitoolkit/projects/clo/housefinchdisease.

Howell, Christine A., William D. Dijak, and Frank R. Thompson. "Landscape Context and Selection for Forest Edge by Breeding Brown-Headed Cowbirds." *Landscape Ecology* 22, no. 2 (2007): 273–284. https://doi.org/10.1007/s10980-006-9022-1.

Irvine, Leslie. *If You Tame Me: Understanding Our Connection with Animals*. Philadelphia: Temple University Press, 2004.

Iwamura, T., H. P. Possingham, I. Chades, C. Minton, N. J. Murray, D. I. Rogers, E. A. Treml, and R. A. Fuller. "Migratory Connectivity Magnifies the Consequences of Habitat Loss from Sea-Level Rise for Shorebird Populations." *Proceedings of the Royal Society B: Biological Sciences* 280 (2013). https://doi.org/10.1098/rspb.2013.0325.

Jasper, James M. *The Art of Moral Protest: Culture, Biography, and Creativity in Social Movements*. Chicago: University of Chicago Press, 1997.

Jasper, James M., and Jane D. Poulsen. "Recruiting Strangers and Friends: Moral Shocks and Social Networks in Animal Rights and Anti-nuclear Protests." *Social Problems* 42, no. 4 (1995): 493–512.

Jerolmack, Colin. *The Global Pigeon*. Chicago: University of Chicago Press, 2013.

Kalof, Linda. *Looking at Animals in Human History*. London: Reaktion Books, 2007.

Kalof, Linda, and Ramona Fruja Amthor. "Cultural Representations of Problem Animals in National Geographic." *Études Rurales* 185 (2010): 165–180.

Kalof, Linda, and Amy Fitzgerald. "Reading the Trophy: Exploring the Display of Dead Animals in Hunting Magazines." *Visual Studies* 18, no. 2 (2003): 112–122. https://doi.org/10.1080/14725860310001631985.

Kalof, Linda, Cameron Whitley, Stephen Vrla, and Jessica Bell Rizzolo. "Anthropogenic Food Sources in the Co-existence of Humans with Liminal Animals in Northern Environments: Representations from National Geographic Magazine." In *Shared Lives of Humans and Animals: Animal Agency in the Global North*, edited by Tuomas Rasanen and Taina Syrjamaa, 147–162. New York: Routledge, 2017.

Kalof, Linda, Joe Zammit-Lucia, and Jennifer Rebecca Kelly. "The Meaning of Animal Portraiture in a Museum Setting: Implications for Conservation." *Organization & Environment* 24, no. 2 (2011): 150–174. https://doi.org/10.1177/1086026611412081.

Kennedy, Emily Huddart. "Rethinking Ecological Citizenship: The Role of Neighbourhood Networks in Cultural Change." *Environmental Politics* 20, no. 6 (2011): 843–860. https://doi.org/10.1080/09644016.2011.617169.

Kennett, Rod. "Participatory Monitoring and Management, and Citizen Science." International Union for Conservation of Nature, 2015. https://www.iucn.org/content/participatory-monitoring-and-management-and-citizen-science.

Kidder, Jeffrey L. "'It's the Job That I Love': Bike Messengers and Edgework." *Sociological Forum* 21, no. 1 (2006): 31–54. https://doi.org/10.1007/s11206-006-9002-x.

Kim, Claire Jean. "Moral Extensionism or Racist Exploitation? The Use of Holocaust and Slavery Analogies in the Animal Liberation Movement." *New Political Science* 33, no. 3 (2011): 311–333. https://doi.org/10.1080/07393148.2011.592021.

Lanham, J. Drew. *The Home Place: Memoirs of a Colored Man's Love Affair with Nature*. Minneapolis, MN: Milkweed Editions, 2016.

Larson, William. "New Estimates of Value of Land of the United States," 2015. https://doi.org/10.1177/0042098014557208.

La Sorte, F. A., D. Fink, P. J. Blancher, A. D. Rodewald, V. R. Gutierrez, K. V. Rosenberg, W. M. Hochachka, P. H. Verburg, and S. Kelling. "Global Change and the Distributional Dynamics of Migratory Bird Populations Wintering in Central America." *Global Change Biology* 23, no. 12 (2017): 5284–96.

Law, Edith, Krzysztof Z. Gajos, Andrea Wiggins, Mary L. Gray, and Alex Williams. "Crowdsourcing as a Tool for Research: Implications of Uncertainty." *Proceedings of the 2017 ACM Conference on Computer Supported Cooperative Work and Social Computing* (2017): 1544–1561, 2017. https://doi.org/10.1145/2998181.2998197.

Lewis, Daniel. *The Feathery Tribe: Robert Ridgway and the Modern Study of Birds*. New Haven, CT: Yale University Press, 2012.

Louder, Matthew I. M., Wendy M. Schelsky, Amber N. Albores, and Jeffrey P. Hoover. "A Generalist Brood Parasite Modifies Use of a Host in Response to Reproductive Success." *Proceedings of the Royal Society B Biological Sciences* 282 (2015): 20151615. https://doi.org/10.1098/rspb.2015.1615.

Martin, Jennifer Adams. "When Sharks (Don't) Attack: Wild Animal Agency in Historical Narratives." *Environmental History* 16, no. 3 (2011): 451–455. https://doi.org/10.1093/envhis/emr051.

McFarlane, Bonita L. "Specialization and Motivations of Birdwatchers." *Wildlife Society Bulletin* 22, no. 3 (1994): 361–370. http://www.jstor.org.

Merriam, Florence A. *Birds through an Opera-Glass*. New York: Houghton Mifflin, Chautauqua Press, 1889.

Mills, C. Wright. *White Collar: The American Middle Classes*. New York: Oxford University Press, 1951.

Mitchell, Charles. "The Bard's Bird; or, The Slings and Arrows of Avicultural Hegemony: A Tragicomedy in Five Acts." In *Trash Animals: How We Live with Nature's Filthy, Feral, Invasive, and Unwanted Species*, edited by Kelsi Nagy and Philip David Johnson II, 171–181. Minneapolis: University of Minnesota Press, 2013.

Mohr, John W. "Measuring Meaning Structures." *Annual Review of Sociology* 24 (1998): 345–370.

Moore, Eric. "Birding: Recent Experiences in Nature Underscore the Importance of Being Observant." *Daily Courier* (Prescott, AZ), September 21, 2017.

Moore, Lisa Jean. *Catch and Release: The Enduring yet Vulnerable Horseshoe Crab*. New York: New York University Press, 2017.

Moore, Lisa Jean, and Mary Kosut. *Buzz: Urban Beekeeping and the Power of the Bee*. New York: New York University Press, 2013.

Mulder, Raoul A., Patrick Jean Guay, Michelle Wilson, and Graeme Coulson. "Citizen Science: Recruiting Residents for Studies of Tagged Urban Wildlife." *Wildlife Research* 37 (2010): 440–446. https://doi.org/10.1071/WR10007.

Nagy, Kelsi, and Philip David Johnson II. "Introduction." In *Trash Animals: How We Live with Nature's Filthy, Feral, Invasive, and Unwanted Species*, edited by Kelsi Nagy and Philip David Johnson II, 1–27. Minneapolis: University of Minnesota Press, 2013.

Nimmo, Richie. "Bovine Mobilities and Vital Movements: Flows of Milk, Mediation and Animal Agency." In *Animal Movements. Moving Animals: Essays on Direction, Velocity and Agency in Humanimal Encounters*, edited by Jacob Bull, 57–74. Uppsala Centre for Gender Research, Sweden: Uppsala University, 2011.

———. *Milk, Modernity and the Making of the Human: Purifying the Social*. London: Routledge, 2010. https://doi.org/10.4324/9780203867334.

Norgaard, Kari Marie. *Living in Denial: Climate Change, Emotions, and Everyday Life*. Cambridge, MA: MIT Press, 2011.

Ottinger, Gwen. "Epistemic Fencelines: Air Monitoring Instruments and Expert-Resident Boundaries." *Philosophy of Science* 3, no. 1 (2009): 55–67. https://doi.org/10.4245/sponge.v3i1.6115.

Peterson, Roger Tory. *Peterson Field Guide to Birds of Eastern and Central North America*. New York: Houghton Mifflin, 2010.

Pimentel, David, Rodolfo Zuniga, and Doug Morrison. "Update on the Environmental and Economic Costs Associated with Alien-Invasive Species in the United States." Special issue, *Ecological Economics* 52, no. 3 (2005): 273–288. https://doi.org/10.1016/j.ecolecon.2004.10.002.

Price, Jennifer. *Flight Maps: Adventures with Nature in Modern America*. New York: Basic Books, 1999.

Ratcliff, Jessica. *The Transit of Venus Enterprise in Victorian Britain*. London: Pickering & Chatto, 2008.

Riesch, Hauke, and Clive Potter. "Citizen Science as Seen by Scientists: Methodological, Epistemological and Ethical Dimensions." *Public Understanding of Science* 23, no. 1 (2014): 107–120. https://doi.org/10.1177/0963662513497324.

Ritvo, Harriet. "Going Forth and Multiplying: Animal Acclimatization and Invasion." *Environmental History* 17, no. 2 (2012): 404–414. https://doi.org/10.1093/envhis/emr155.

Robinson, John C. "Relative Prevalence of African Americans among Bird Watchers," USDA Forest Service, 2005.

Runge, Claire A., James E. M. Watson, Stuart H. M. Butchart, Jeffrey O. Hanson, Hugh P. Possingham, and Richard A. Fuller. "Protected Areas and Global Conservation of Migratory Birds." *Science* 350, no. 6265 (2015): 1255–1258. https://doi.org/10.1126/science.aac9180.

Russell, Joshua. "'Everything Has to Die One Day': Children's Explorations of the Meanings of Death in Human-Animal-Nature Relationships." *Environmental Education Research* 23, no. 1 (2017): 75–90. https://doi.org/10.1080/13504622.2016.1144175.

Schaffner, Spencer. *Binocular Vision: The Politics of Representation in Birdwatching Field Guides*. Amherst: University of Massachusetts Press, 2011.

———. "Environmental Sporting: Birding at Superfund Sites, Landfills, and Sewage Ponds." *Journal of Sport & Social Issues* 33, no. 3 (2009): 206–229. https://doi.org/10.1177/0193723509338862.

Scott, David, and Jack Thigpen. "Understanding the Birder as Tourist: Segmenting Visitors to the Texas Hummer/Bird Celebration." *Human Dimensions of Wildlife* 8 (2003): 199–218. https://doi.org/10.1080/10871200390215579.

Scroggins, Michael. "Ignoring Ignorance: Notes on Pedagogical Relationships in Citizen Science." *Engaging Science, Technology, and Society* 3 (2017): 206–223. https://doi.org/10.17351/ests2017.54.

Shapiro, Kenneth J. "Animal Rights versus Humanism: The Charge of Speciesism." *Journal of Humanistic Psychology* 30, no. 2 (1990): 9–37.

Shepherd, Hana. "Classification, Cognition and Context: The Case of the World Bank." *Poetics* 38, no. 2 (2010): 133–149. https://doi.org/10.1016/j.poetic.2009.11.006.

Shirk, Jennifer. "Push the Edge of Science Forward: Expanding Considerations of Expertise through Scientists' Citizen Science Work in Conservation." PhD diss., Cornell University, Ithaca, NY, 2014.

Simmel, Georg. "The Metropolis and Mental Life." In *The Sociology of Georg Simmel*, 409–424. New York: Free Press, 1950.

Snyder, Gary. *The Practice of the Wild*. New York: North Point Press, 1990.

Steinfeld, Henning, Pierre Gerber, Tom Wassenaar, Vincent Castel, Mauricio Rosales, and Cees De Haan. "Livestock's Long Shadow: Environmental Issues and Options." *Food and Agriculture Organization of the United Nations*, 2006. http://www.fao.org/docrep/010/a0701e/a0701e00.HTM.

Stranger, Mark. *Surfing Life: Surface, Substructure and the Commodification of the Sublime*. London: Routledge, 2017.

Strycker, Noah. *Birding without Borders: An Obsession, a Quest, and the Biggest Year in the World*. New York: Houghton Mifflin, 2017.

Sullivan, Brian L., Christopher L. Wood, Marshall J. Iliff, Rick E. Bonney, Daniel Fink, and Steve Kelling. "eBird: A Citizen-Based Bird Observation Network in the Biological Sciences." *Biological Conservation* 142 (2009): 2282–92. https://doi.org/10.1016/j.biocon.2009.05.006.

Thornton, Sarah. *Club Cultures: Music, Media, and Subcultural Capital*. Middletown, CT: Wesleyan University Press, 1996.

Tidwell, Mike. "The Low-Carbon Diet." *Audubon*, January-February 2009. http://www.audubon.org/magazine/january-february-2009/the-low-carbon-diet.

Tovey, Hilary. "Theorising Nature and Society in Sociology: The Invisibility of Animals." *Sociologia Ruralis* 43, no. 3 (2003): 196–215.

Twining, Hillary, Gary Patronek, and Arnold Arluke. "Managing the Stigma of Outlaw Breeds: A Case Study of Pit Bull Owners." *Society and Animals* 8, no. 1 (2000): 25–52. https://doi.org/10.1163/156853000510970.

Turner, Jack. *The Abstract Wild*. Tucson: University of Arizona Press, 1996.

Urry, John. *The Tourist Gaze: Leisure and Travel in Contemporary Societies*. London: Sage, 1990.

U.S. Fish and Wildlife Service. "2011 National Survey of Fishing, Hunting, and Wildlife-Associated Recreation," 2011. https://doi.org/10.3886/ICPSR34699.

———. "Birding in the United States: A Demographic and Economic Analysis," 2013. https://digitalmedia.fws.gov/cdm/ref/collection/document/id/1874.

Van Dooren, Thom. *Flight Ways: Life and Loss at the Edge of Extinction*. New York: Columbia University Press, 2014.

Vincent, Carol Hardy, Laura A. Hanson, and Jerome P. Bjelopera. "Federal Land Ownership: Overview and Data." *Congressional Research Service*, 2014, 28. https://www.fas.org/sgp/crs/misc/R42346.pdf.

Weidensaul, Scott. *Of a Feather: A Brief History of American Birding*. Orlando, FL: Harcourt, 2007.

Wiggins, Andrea, and Kevin Crowston. "From Conservation to Crowdsourcing: A Typology of Citizen Science." In *HICSS '11 Proceedings of the 2011 44th Hawaii International Conference on System Sciences*, 1–10. Kauai, HI: IEEE, 2011. https://doi.org/10.1109/HICSS.2011.207.

Wiggins, Andrea, Greg Newman, Robert D. Stevenson, and Kevin Crowston. "Mechanisms for Data Quality and Validation in Citizen Science." In *Proceedings of the 2011 IEEE Seventh International Conference on E-Science Workshops*. Stockholm, Sweden: IEEE, 2011. https://doi.org/10.1109/eScienceW.2011.27.

Wohlsen, Marcus. *Biopunk: Solving Biotech's Biggest Problems in Kitchens and Garages*. New York: Penguin, 2012.

Wu, Joanna X., Chad B. Wilsey, Lotem Taylor, and Gregor W. Schuurman. "Projected Avifaunal Responses to Climate Change across the U.S. National Park System." *Plos One* 13, no. 3 (2018): e0190557. https://doi.org/10.1371/journal.pone.0190557.

Wynes, Seth, and Kimberly A. Nicholas. "The Climate Mitigation Gap: Education and Government Recommendations Miss the Most Effective Individual Actions." *Environmental Research Letters* 12 (2017): 1–9. https://doi.org/10.1088/1748-9326/aa7541.

Wynne, Brian. "Misunderstood Misunderstanding: Social Identities and Public Uptake of Science." *Public Understandings of Science* 1 (1992): 281–304.

Zerubavel, Eviatar. *The Fine Line: Making Distinctions in Everyday Life*. New York: Free Press, 1991.

———. *Hidden in Plain Sight: The Social Structure of Irrelevance*. New York: Oxford University Press, 2015.

INDEX

Note: Pages in *italics* refer to illustrations.

ABOUT THE AUTHOR

ELIZABETH CHERRY is an associate professor of sociology at Manhattanville College in Purchase, New York. She is the author of *Culture and Activism: Animal Rights in France and the United States*.